Medicinal Plants, Forests and Their Conservation

Medicinal Plants, Forests and Their Conservation

Ramesh Bhadari

CYBER TECH PUBLICATIONS
4264/3, Ansari Road, Daryaganj, New Delhi-110002 (India)
Ph.: 011-23244078, 011-43559448 Fax: 011-23280028
E-Mail: cyberpublicationsdelhi@yahoo.com
Website: www.cybertechpublications.com

MEDICINAL PLANTS, FORESTS AND THEIR CONSERVATION

Ramesh Bhadari

First Edition 2011

Published by :

G.S. Rawat for **Cyber Tech Publications**
4264/3, Ansari Road, Daryaganj, New Delhi-110002 (India)
Ph.: 011-23244078, 011-43559448 Fax: 011-23280028
E-Mail: cyberpublicationsdelhi@yahoo.com
Website: www.cybertechpublications.com

CONTENTS

Preface

A major event was the Eighth World Forestry Congress, devoted to the theme " Forests for People" that gave a thrust to community forestry. In an address to Congress, even Westoby took a major departure from his previous stance, acknowledging, "the dreamed snowball-effect of forest industries on rural economies has not materialised. The term social forestry first surfaced in 1976 report of the National Commission of agriculture in India, in which it was used for a programme of activities to encourage those who depend on fuelwood and other forest products to produce their own supplies– in order to "lighten the burden on production forestry" (GOI, 1976 as cited by Arnold, 1991).

A shift from 'Keynesian style' development paradigm to a more rural oriented development approach paved way for the emergence of Social Forestry. In the early 1950's the role of forests was to provide raw materials to the forest industries. This approach was based on a high " structural interdependence between forestry and the industrial sector of the economy (Westoby, 1962). While industrial forestry was at the heel, E. Eckholm pointed out in 'The other energy crisis: firewood' that " for more than a third of world's population, the real energy crisis is a daily scramble to cook dinner" (Eckholm, 1975).

Hence, In mid 1970s there were shifts in the developmental thinking. This shift stemmed from the philosophy that the development should be achieved " based on rural income and output". The importance was placed on achievements of equity, emphasizing the distribution aspects underlying growth. Also people's participation in the development process was stressed.

—Author

1

Introduction

Social Forestry

The term social forestry first surfaced in 1976 report of the National Commission of agriculture in India, in which it was used for a programme of activities to encourage those who depend on fuelwood and other forest products to produce their own supplies– in order to "lighten the burden on production forestry" (GOI, 1976 as cited by Arnold, 1991).

A shift from 'Keynesian style' development paradigm to a more rural oriented development approach paved way for the emergence of Social Forestry. In the early 1950's the role of forests was to provide raw materials to the forest industries. This approach was based on a high " structural interdependence between forestry and the industrial sector of the economy (Westoby, 1962). While industrial forestry was at the heel, E. Eckholm pointed out in 'The other energy crisis: firewood' that " for more than a third of world's population, the real energy crisis is a daily scramble to cook dinner" (Eckholm, 1975).

Hence, In mid 1970s there were shifts in the developmental thinking. This shift stemmed from the philosophy that the development should be achieved " based on rural income and output". The importance was placed on achievements of equity, emphasizing the distribution aspects underlying growth. Also people's participation in the development process was stressed. A major event was the Eighth World Forestry Congress, devoted to the theme " Forests for People" that gave a thrust to community forestry. In an address to Congress, even Westoby took a major departure from his previous stance, acknowledging, "the dreamed snowball-effect of forest industries on rural economies has not materialised" (Westoby, 1987). Westoby later acknowledged: *"In the early......I had concluded that forestry is about trees. But of course, this is quite wrong. Forestry is not about trees, it is about people. And it is about trees only insofar as trees can serve the needs of people."* (Leslie, 1987).

Hence, development thinking, and practice, was therefore moving towards a rural led focus and the need to help rural populations mobilised by devoting greater efforts towards meeting their "basic needs" – a shift which took concrete form in the World Conference on Agrarian Reform and Rural Development (WCARRD) held by FAO in July 1979.

In 1978 FAO and SIDA launched a special action programme to heighten the importance of "community forestry" after the publication of " Forestry for local community development" (FAO, 1978).

Also in 1978, the World Bank issued Forestry Sector Policy with a major shift from industrial forestry towards environmental protection and meeting the local needs (World Bank, 1978). Another initiative was the creation of ICRAF in 1977 by IDRC for promoting research in " Agro-forestry" (Bene et. al., 1977).

CONCEPTS AND DIFFERENT PRACTICES OF SOCIAL FORESTRY

Social forestry refers to a group of management strategies (Agro-forestry farm forestry, community forestry and publicly managed forestry) for local community development, in which the aspects of local participation in management and in benefits of tree growing are central objectives (Wiersum, 1991).

Important objectives of Social Forestry are as follows:

1. Reduction of environmental degradation.
2. Increasing supply of tree/forest products.
3. Increasing income and employment.

The above objectives are interdependent – not only technically but achieving an objective is often a pre-requisite from socio-economic point of view e.g., reduction of environment degradation will not be achieved unless income objective is reached.

Agro-forestry

Agro-forestry is a technique often used in social forestry that attempts to accommodate farmer's objective in relationship with tree growing. It is usually defined as the deliberate combination of wood perennials (trees, shrubs, bamboo etc.) on the same land management unit as agriculture crops, pastures and animals. This combination may be obtained in a mixed spatial arrangement in the same place at the same time or a temporal sequence.

There are different systems of Agro-forestry and their interactions with other crops is summarised in table-1.

Farm Forestry

Farm forestry is the deliberate attempt of farmers towards planting or keeping trees in their farmland in order to supplement daily farm/household needs, meet the contingencies and take market opportunities created by the dwindling supply of tree/ forest products from the common lands.

There are reports on trees being increasingly used as contingencies and also contributing to the household food security. Besides, the above points, the tress also contributes to the sustainable land use.

Farm forestry: Response towards the scarcity

The real crisis faced by the Third World Countries where firewood, crop residues, kerosene being the popular form of energy for cooking, Eckholm wrote "the people are scrambling to cook dinner under deforestation" (Eckholm, 1975). However, in the wake of deforestation, villagers coped with increased scarcity in various ways (Ghimere, 1994). Introduction of tress in the private land is widely pursued by the farmers elsewhere. A study, using aerial photographs between 1964 and 1988, observed a tremendous increase of private trees in the hills of Nepal (Carter and Gilmour, 1989). It is postulated that this increase is mainly due to the decreasing access and unavailability of forest resources (Gilmour, 1990). This shift of trees has occurred elsewhere. Pakistan gives example of Hazara forests managed by villagers in frontier tribal areas under weak state control (Dove, 1995). An African study shows " rural households are planting trees in the wake of labour shortages due to off-farm income opportunities.

Farmers favour multipurpose trees over trees that supply only firewood, as cheap substitutes for firewood are available. Similarly planting trees only for fodder alone is not common, but existing fodder trees are retained (Werner, 1995, Dewees, 1995). Thus introduction of trees in private land is a strategy adopted by villagers to cope with emerging biomass scarcity. Timber trees are stores of values for household savings that do not put demand on household's scarce cash resources (Chambers and Leach, 1989). However, tree growing is determined by the farmer's livelihood strategies and resource base. Farmer's decision regarding tree growing depends on:

(1) declining access to the wood land,

(2) loss of communal lands or restrictions on forest access

(3) increasing demand for forest products

(4) increasing demand from the market.

A survey from western Kenya shows that the "poor" prefer firewood. The "average" had a higher portion of fruit and timber while the wealthy invests heavily in fencing (Scherr, 1995). A study on farmer's willingness to grow trees in Gunung Kidul opts for a different reason. As the farmer's willingness to grow trees depends on many factors. An increase in the productivity of staple crop is an important factor permitting farmers to plant trees. Government policy can also create a favourable market trend for growing trees (Filius, 1997)

Constrains of tree growing

The following are the main constraints for tree growing by the rural poor (FAO, 1985):

1. Economic: trees can't compete with other uses for land, labour and capital.
2. Social/cultural: shelter to evil spirits, harbours insects etc.
3. Site: condition not meeting requirement for trees
4. Land use rights tenure security
5. Red tape(ism): Permits for utilisation

The above constraints affect the farmer's willingness to grow trees while Gregersen et al., 1989 gives some reasons with regard to flexibility that can be (dis)incentive for farmers to grow trees:

- Multiple tree outputs (adjust as the market change)

- Flexibility in harvesting time
- Low liquidity in early years reduces flexibility.

Besides the above, the higher rate of discount in case of developing countries discouraged farmers from introducing trees in their farmlands.

Social forestry interventions in south Asia intended to involve local communities, whether it is agro forestry, farm forestry or community forestry needs well planned transformation of a dynamic inter-relationship among community, natural environment and the state (Agarwal 1995). In certain situations, the forest-community dependencies are deep rooted and need careful consideration while designing such development interventions. The theoretical basis for planning such transformations is more complex than project planners and bureaucrats often consider. Most of these forestry interventions (for the purpose of this paper termed as 'social forestry') went through a number of paradigm shifts, beginning with a top down conventional model approach, to the most recent approach wherein emphasis is laid on local participation. There is a lack of proper understanding by policy makers and foresters, and inadequate theoretical basis in these approaches. Analysis of the approaches may provide explanation for the poor performance of most social forestry projects in south Asia. Different views exist about the poor outcomes and the basic reasons for such outcomes.

Earlier studies that addressed issues of social forestry in south Asia tended to neglect local conceptualisations of natural resources, because their methodologies excluded local histories, local socio-political process and issues of institutional change. Studies in the recent years have started paying attention towards the analysis of local livelihoods, the local complexity of natural resource management and the long traditions of local resource management and conservation (Blaikie and Brookfield 1999; IIED 1998; IDS 1999).

This study will attempt to analyse some of the theoretical contents of the social forestry projects in south Asia and related policies in order to make them explicit and the factors responsible for the poor performance of the social forestry projects with regard to equity, environmental rehabilitation and sustainability.

The present study draws on the authors' research and field experience in social forestry in South Asia (India, Nepal and Sri Lanka) since the mid 90s. Some data was also gathered during specific reviews of three social forestry projects in India, Nepal and Sri Lanka funded by different international donor agencies. The reviews consisted of discussions with local key informants and government officials, several field visits and secondary data collection.

DEFORESTATION CRISIS AND SOCIAL FORESTRY

Generalizations about deforestation have been termed as "received wisdom" (Leach and Mearns, 1996), "orthodoxies" (Benjaminsen 1993, 1998), "mainstream views" (Sandford, 1983), "narratives" (Roe, 1991). Studies supporting these mainstream views on deforestation are built on the assumptions that use of fuelwood, fodder and other forest products by local people are the most important factors leading to forest degradation. This view has been presented in a number of publications (IUCN *et al.* 1980; Anderson and Fishwick, 1984; Timberlake, 1985; WCED, 1987). In some cases the situations were presented as serious, and produced an immediate need for enhancement of biomass supply to mitigate the fuelwood and fooder crisis. The mainstream views, encouraged donors and national governments to go for large-scale afforestation projects in many regions including south Asia. Social forestry was originally conceived by the national governments as a response to the 'fuelwood crisis' and to accelerating deforestation. But, deforestation is more complex than suggested by the population-based models. The deforestation theories blaming locals were promoted by stakeholders with vested interests. For example, the Fuel Policy Committee set up by Government of India in 1974 noted that:

The recorded fuelwood output (from Indian Forests) in 1970 is about nine million tonnes. The actual consumption of firewood is, however reported to be over 100 million tonnes, the balance coming from unrecorded felling from forests (GOI, 1976, pp. 23).

The report issued a warning that; if fuelwood extraction would continue in the same manner it will lead to complete depletion of forest cover within 10-20 years. Similarly, the World Bank review of the Nepal forestry sector in 1978 resulted in an alarmist report premised on the 'theory of environmental degradation'. Such debates are relevant not to just explain past deforestation, but also have implications on planning strategies to solve the crisis and prevent recurrences in the future. The solutions would have been different if the debate had taken a broader perspective emphasizing on other factors responsible for deforestation, for example, agricultural expansion, timber extraction by local or multinational companies, mining, weaknesses in policy and legislation, weak institutions, corruption etc, rather than just the unsustainable exploitation by local people. In the latter case, the problem lies with the local people, and in the former it is do more with the broader society. Consequently, the solution would also depend on where the 'problematique' is located. Agarwal (1998) highlights the government's interests for acquiring funds from donor agencies, behind promoting these deforestation discourses and how the forestry projects during 80s were based on such discourses. In India, the National Commission on Agriculture (NCA) in 1976 recommended the creation of 'social forests' on common lands and state forests to provide the local communities fuel wood, small timber and fodder (GOI, 1976).

The following project background description by an external consultant for Department of International Development (United Kingdom) in 1987, shows how, even the donor agencies and national governments justified the social forestry projects based on the 'Fuelwood Orthodoxy' approach:

"India's designated forests cover about 75 million ha, but increasing pressures from local human and cattle populations have resulted in severe depletion of much of the natural forests. Fuelwood, extraction from forests is leading to widespread deforestation. A massive increase in afforestation is needed to ensure adequate supplies of fuelwood and fodder, building poles and other products for the rural poor. To achieve this the development of forest resources through social forestry is essential."

As a consequence, social forestry schemes by and large remained technocratic throughout the sub-continent emphasizing on biomass production and failed to respond to the needs of the poor and landless (FAO, 1993). However, in recent years there has been a growing awareness and research investigating the mainstream views in local settings (Benjaminsen, 1993; Fairhead and Leach, 1996; Agarwal, 1998; Udaya Sekhar 2000). These studies are providing more realistic assessments of deforestation and its causes, and giving a more respectful attention to local knowledge and practices that are the necessary bases for effective and appropriate environmental policies. According to them, deforestation is not merely a result of gradual local pressure on resources, but more of disruption of the traditional institutional framework responsible for resource protection and management. These studies demonstrate that extraction of wood by local communities is not the main reason for deforestation and that exclusion of local people does not help to realize the objectives of the studied projects.

LOCAL PEOPLE AND NATURAL RESOURCES MANAGEMENT

Conservation and management of biomass resources and environment are embedded in an array of knowledge and local practices. Most development projects tend to ignore the ingenuity of rural people in responding to the local problems based on their local knowledge (Chambers, 1983). Initiatives intended to grow trees through Social Forestry projects have not taken cognisance of the existing local practises. For example in the Himalayan region, common practices of obtaining biomass resources from forests are through pollarding, pruning or selective cutting and not by felling of tress (Saethre, 1993). Similarly studies from Nepal have shown that, communities adjacent to forests mostly use dry wood as fuelwood and grasses for fodder that normally does not involve cutting of trees (Sharma and Shaw, 1993). For many local communities across South Asia, forests are not only of consumptive value, but also revered as places of great socio-cultural and religious significance (Chandrakanth *et al.*; 1990; Udaya Sekhar, 2001). The other factors that regulate resource extraction from forests include (a) traditional rights (b) religious practices (c) and non-use of certain tree species in the region. For example, people have a strong belief that felling of certain tree species such as 'neem' (*Azadirachta indica*) and 'pipal' (*Ficus religiosa*) brings evil to the members of the household. Thus local systems of knowledge and management are sometimes rooted deeply in their religion and belief systems, which help in conserving the resources. But, state management was often based on flawed premises or a lack of understanding of the people-nature dependency (Chandrakanth *et al.*, 1990; Sharma, 1991). Gradually local communities have lost access and control of the resources, resulting in conflicts with the state and other stakeholders.

In many parts of south Asia, local people maintained strict rules to manage natural forests in their vicinity even until the 60s (Jodha, 1992; Agarwal and Yadama, 1997; Udaya Sekhar,

2000). The examples given below are illustrative of the diversity of informal or traditional institutional arrangements at the local level:

(i) A grazing fee was levied on herders in some pasturelands. In some regions, grazing was banned in pond catchments to prevent siltation of ponds.

(ii) Cutting of grass and fodder leaves from the certain tree species, 'Sal' (*Shorea robusta*) in the *Terai* region in the Himalayas, or 'Palas' (*Butea monosprema.*) in central and western India was regulated on forestlands.

(iii) Collection of dry wood did not require individual permits from the informal councils at the village level, whereas special permits were required for felling trees on the hill slopes and catchments. In case of some species that are less in abundance such as Shisham (*Dalbergia sisoo*) and Bamboo (*Dendrocalamus* spp.) there was a ban on felling. These species had market value for their timber.

The village community performed collective work to enhance and maintain the tree cover, by fencing, planting, trenching in forests, fire lines maintenance and desilting of ponds. Each household had an obligation to contribute a share to such efforts, failing this its rights to resources were curtailed. These duties are no longer followed as the local rights to resources are curtailed with state control.

The indirect non-consumptive values were represented in day-to-day life in the form of rules and conventions regulating use of resources. The informal institutions tend to reduce the environmental uncertainty faced by resource users especially in arid and semi-arid regions and also contribute to the conservation of natural resources (Agarwal and Yadama 1997; Jodha 1993; Runge 1992). In some cases, the local institutions are better informed than the state departments about local ecological, economical and social conditions and about problems and constraints, which may be important from the management point of view. According to some studies, traditional management systems of CPRs have contributed to the protection of natural resources from being over used and have played a vital role in the management and conservation of natural resources (Cordell & Mckean 1990; Feeny *et.al.*,1990).

TENURE RIGHTS AND BIOMASS RESOURCES

Prior to 1950s, local landlords controlled and had the authority to grant rights in forests in India, Sri Lanka and Nepal. In different areas, the tenure systems varied, for example the *Birta* system in mid hills of Nepal and the *Zamindari* system in northern part of India. The colonial regime in India and Sri Lanka significantly influenced the land tenure system. Under colonial management systems which continued in some areas even during the postcolonial period, the local landlords had the responsibility to manage forests and granted rights to the local households. These include access rights, use rights and entitlement rights, which determine the access and end use of resources from the forests. Timber extraction in general was regulated, but there was free access to non-timber forest products. These resources were treated by local communities in many parts of south Asia as a "free good" which were easily available. But these rights and duties ceased with the land reforms during 50s and subsequent development policies initiated by respective governments. In Nepal, forests were nationalized (under the Private Forests Nationalization Act of 1957 and the Birta Abolition Act of 1959) as a part of a wider move to break feudal structures. In India, it was the Land Revenue Act in the respective states, which was implemented to nationalize forests. Nationalization of forests did more harm to landless and rural poor whose access to common pool resources was restricted as part of these resources were declared protected.

In addition, communal sources of fuel wood, fodder and non-timber products have been reduced as a result of the forestry projects under the management and control of state functionaries. Due to nationalization of forest, the lack of local tenure and rights meant there was no incentive for local users, especially the poor and landless to exercise restraint. The rights to access and use of forest resources was further restricted with social forestry interventions enclosing forestlands. For example, in the middle hill region of Nepal, the poorer sections of the people have less access to forest products for subsistence use and income today, than they had before the implementation of social forestry projects (Hobley, 1996). This is also the case in India and Sri Lanka, where common lands, which were accessible to the landless prior to the social forestry intervention, are no longer freely available (Jodha, 1992). In such a situation, the cultural overlay of rights creates paradoxically huge 'scarcity' differentials between households in villages, especially for the landless and poor. This tends to inhibit effective development and active participation by marginalized groups. Currently, the average poor household receives only one third to one fifth of the wood from forests than collected before the social forestry intervention (Agarwal 1998). The rights to use resources from forestlands, which were formerly guaranteed by customary law, were banned after the land reforms and land acts came into existence in 1950s.

The choice of species or location of social forestry plantations was often made by people who had practically little local knowledge. The fact that there are local variations, even in small countries like Sri Lanka and Nepal, wherein women traditionally prefer certain tree species for fodder or fuel wood is important for any afforestation project to consider. The practice of use restrictions and the issue of land tenure, when viewed in the context of cultural and property rights based on gender inequalities and power relations within the households, presents a complex scenario. The tenure rights to land and resources in many South Asian regions are held by men, giving them legal rights over the resources, although women perform the frequent work of collecting biomass resources for household needs, cooking etc. Attached daily to the forest, women thus have a better knowledge on forest management and conservation, but are the most neglected group in social forestry projects (Chandrakanth et al., 1990).

There is no equity between men and women in land tenure specified officially in the land laws of the respective countries. Neither are women's rights in resource use are ensured in practice. Local traditions seems strongly influence on the women rights in land tenure, as it is the male child who inherits land. As a consequence, women's involvement in forestry projects is less, at the most they are being used as labourers in planting or watering the plants.

SOCIAL FORESTRY INTERVENTION

Social forestry is any practice, method, technique/technology or natural resource management system that enhances forest resource governance and makes forestry economically viable, and ecologically sound (GOI 1976). This definition is mostly a state view of what social forestry should be, and does not emphasise the involvement of local people. It advocates that forest plantations should be controlled by the state within fences, thus excluding local people and livestock from entering it. Projects designed with such an approach, for example the National Social Forestry Project (NSFP), aided by the World Bank in India, are typical examples dominated by the orthodox views of deforestation. In general, social forestry interventions initially emerged from concerns over forest degradation as discussed earlier. These forestry interventions, whether it was community forestry (CF) in Nepal or social forestry in India and Sri Lanka, had their focus on technical questions of resource management. It was only towards the beginning of 1990s that the shortcomings of the approaches became evident and since then social forestry interventions started taking a new direction. The forestry interventions started to encompass a rhetoric on a more 'people-oriented approach'. Social forestry or CF was then redefined as a concept, "where forest is managed by local people (living near or in the concerned forest) for their own benefit". However, the new approach still does not explicitly address livelihoods and local resource management patterns. The fundamental concept of CF in Nepal is that local forest users are given collective management responsibility for the local forests on which they depend for product flows. The model is an extremely defined form of organization that is defined in relation to the resource and not to existing social relations or community structures. The new policies still emphasize more on protection and regeneration of the resource, in response to the heightened concerns over rapidly deteriorating forest resources. Under the new management approach, the Joint Forest Management (JFM) committees in India or the Forest User Groups (FUGs) in Nepal find it more difficult to operate in some situations, as the state agencies become more strict as these are supported by policy amendments. For example, the Forest (Second Amendments) Bill, 2001 in Nepal amended the 1993 Forest Act in a number of fundamental ways, which have restricted the FUGs. Under this amendment, the FUG revenues will be taxed at 65 per cent, and CF in the *Terai* area would be subject to separate and more stringent regulations. This demonstrates the dual nature of the state trying to promote local participation on one hand, and at the same time giving more powers to itself to retain control of the forestland.

In general, justifying social forestry intervention was much easier for government agencies in situations where deforestation crisis was projected as more serious, than in situations where it was considered less serious. In the latter case, it can be difficult for forest agencies to justify their intervention and secure funding. On the other hand, the cost-benefit ratio had to be projected to convince donors about the economic viability of the projects. This led to the recommendation of fast growing species in social forestry plantations, irrespective of local factors and needs. Eucalyptus was one such species recommended by forest planners on a large scale during the 70s and 80s in social forestry projects. Large areas were brought under monocultures of Eucalyptus, which became a major environmental controversy towards the 80s and 90s. In the Terai region of Nepal and India, the Forest agencies, replanted degraded *sal* (Shorea Robusta) forests with exotic tree species, mainly *Eucalyptus* and *Sissoo* (Dalbergia Sissoo), under the pretext that it would produce biomass in short periods. There was resistance from local NGOs opposing the replanting program of the Forest Department, in some cases leading to violent conflicts with local people.

Social Forestry has been part of the national forest policies in South Asian region in one form or another since the 70s. But after three decades of development and implementation, it has still not met its stated objective of promoting plantation activities, which address the needs of poor people for fuel, fodder, small timber and non-timber forest products. According to Agarwal (1995), social forestry has the potential for substantial welfare improvements for the target population, especially women, through increased biomass availability, decreased time for collection and decreased pressure on natural forests. Intervention needs, however, to be very selective to be successful; they must be locally specific with special consideration given to the needs of local people. The goal of social forestry intervention should not only be to provide biomass but also to reverse the processes of negative change through active participation.

SOCIAL FORESTRY IN SOUTH ASIA (REVIEW OF CASE STUDIES FROM INDIA, SRI LANKA AND NEPAL)

Shivapuri Watershed Management and Fuelwood Plantation Project in Nepal (funded by Norway through a trust fund to FAO)

The Shivapuri project was initiated in 1975 by the government of Nepal in order to protect the Shivapuri range as a water catchment area and a wildlife reserve. FAO became involved in the project from 1985, introducing ideas of social forestry in the form of block plantations of fulewood, plantation on private land and other community development activities. The objectives of the project included protection of the watershed, reduction in land degradation and community development. One of the objectives related to biomass is phrased as follows:

To alleviate the heavy demand for fuelwood/fodder in the watershed area by establishment of fuelwood/fodder plantations on available land in and adjoining the project area through community participation to meet local requirements; as well as encouraging the use of alternative energy where practicable.

This objective reflects the standard thinking of planners; keeping people out of forest areas while seeking to develop alternative resources on under-utilised land. The perception was that the degradation of the watershed area is caused by over-utilisation by local people, and that creation of alternative forest resources would reduce the pressure on the protected area.

The government's take over of ownership to the forestlands in 1957 caused a disruption in the traditional ways of land management and governance, and no alternative system was effectively developed in its place. This resulted in indiscriminate extraction of resources encouraged by a high demand for wood products from the nearby capital city of Kathmandu. Also other resources like grass and medicinal plants were collected for sale. As the sale was unofficial, it was difficult to develop a system ensuring a balanced extraction of resources. Thus, the commercial use of the watershed area and the lack of a management system and tenure arrangements were important factors in land degradation.

The unsustainable utilisation of the watershed was mostly by people who did not necessarily need or benefit from fuelwood plantations in the project area. Some were businessmen or farmers with large land holdings, and were not as such in short supply of fuelwood. In addition, the allocation of "available lands" land to community groups also deprived access of some vulnerable groups who had developed survival strategies based on grazing or harvesting grass etc. in these lands.

A part of the degradation in the Shivapuri range was caused by pilgrims from outside visiting the sacred source of the Bagmati river near the Shivapuri summit. In this case the sacredness had an adverse effect caused by fuelwood cutting and wildfires caused by pilgrims.

The expressed need of many families during a local survey was primarily school facilities, drinking water and alternative sources of income and not fuelwood. Although forestry activities had substantial support in the communities, the survey gives no evidence of a fuelwood shortage. Forest extraction may have been of more importance as an income generating activity, and as a source of fodder and building materials etc.

All of these indicate that the strong focus on fuelwood that was seen in the first phase of the project was a too simple approach to a complex problem embedded in social, historical and political factors. The plantation efforts were not necessarily wrong, but it was an incomplete response. Other needs were partially captured by other elements of the project objectives covering community development activities, but these received less focus in the early stages of the project. Such additional strategies were strengthened in the second phase of the project starting in 1991. In this phase the name of the project was changed, removing the fuelwood aspect and naming it the "Shivapuri integrated watershed management project" (FAO 1991). This reflects a change in development thinking. A more varied response to poverty alleviation and land degradation is reflected in the objectives of this phase. From focussing on the physical resources and the physical reasons for their degradation, a shift can be seen in the direction of focusing on the people and their livelihood, and on their use of the resources. The project was still, however, a *project*, not yet encompassing new ideas of a programmatic approach linking the management of the Shivapuri protected area and the development activities in the surrounding villages.

National Social Forestry Project in India (financed by the World Bank)

In India, social forestry programmes began on a large scale only in 1974 in response to the interim report of the National Commission on Agriculture. Social forestry was originally conceived by the Indian government as a response to the forestry crisis and to accelerating deforestation in India. As a consequence, the original objectives of social forestry projects including those financed by the World Bank were to produce fodder, fuel wood, small timber, fruits, and minor forest produce in fenced plantations using fast growing species. In 1984, the World Bank approved the India National Social Forestry project for $165 million. A review of the project document reveals the main objectives as follows:

- To increase production of fuel wood, small timber, poles and fodder.
- To increase rural employment, farmer's incomes and opportunities for participation by landless people.
- To increase the forest cover
- To strengthen forestry institutions.

The objectives show that emphasis is mostly on resource creation through planting fast-growing tree species on privately owned large farms and wastelands. But the products from social forestry plantations were found to have more value for paper and pulp industries and commercial institutions, rather than meeting biomass needs to local people. For example, Eucalyptus, which was the most planted species in the NSFP, was inappropriate in meeting the project's objectives because of its lack of fodder value and ineffectiveness in soil building. Besides, it was inappropriately used as a monocrop in semi-arid areas where competition for water and the need for soil enhancing treatments are high. Several studies expressed that widespread planting of eucalyptus in ecologically inappropriate arid areas has boomeranged with degradation of soils and water tables. An USAID review in 1988 noted that the project was not meeting the subsistence needs of the local people, and therefore was ineffective in reducing pressure on existing forestlands. The opportunities for participation by landless people meant using landless people as daily wage labourers for planting activities. The local participation was thus measured in terms of man-days as labourers, also seen by the donor agencies as a progress indicator.

Gradually, social forestry projects have come under increasing criticism because they have failed to actively involve the local communities and rural poor, who are supposedly the main beneficiaries. Private farmlands, wastelands and community lands have been converted for commercial uses, and in a number of cases the access of poorer rural populations to fodder, fuel wood and other forest products has actually been reduced. The lack of participation of local communities in project design and implementation, and over reliance on industry-biased State forestry departments appear to be the main causes of the project's shortcomings. For example, in Karnataka state of India, small farmers and rural poor were driven to protesting industry biased plantations, including civil disobedience involving the uprooting of seedlings on village common lands.

Similarly, in Bastar region of Madhya Pradesh, which is predominantly a tribal district, the social forestry project involved land covered by natural forests, which was used to raise commercial plantations. The area was planted with tropical pines and other non native species and the biomass produced was not much useful to the tribals whose livelihoods are solely dependant on natural forests. The tribal sustenance base in cane and bamboo for basket weaving, mangoes, tamarind, jackfruit, mahua and edible berries was disrupted when monoculture plantations of eucalyptus or tropical pine replaced natural forests. The project was finally shelved due to the serious resistance of local tribals. It was based not on local knowledge but on ecological ignorance of the forest ecosystem and tribals integration. Such setbacks led to critical reviews both in academic and executive circles, resulting in a much broader approach to address the biomass needs of the poor. In recent years, there is a shift in the terminology, which appears more social, and response from the people to participate in the development process. The solutions to long perceived biomass scarcity in India lie not in the effectiveness of state initiatives in forestry development, but in promoting local forestry management practises through the new Joint forest management policy. However, it is important that the new management paradigm is committed to actively involve people, and ensure equitable distribution of resources, especially to the marginal groups. One of the problems identified in the new paradigm is the retention of the power by the state to make decisions related to sharing of forest resources, membership in the management committees and other day to day activities of the committees which can be left to the people to decide.

The Upper Watershed Management Project, Sri Lanka (Funded by an ADB credit)

The Upper Watershed Management Project (UWMP) objectives were to rehabilitate and sustainably manage and protect the critical watersheds, to improve the income of project beneficiaries and to facilitate the establishment of a medium to long-term watershed management policy. The project has three main components:

- Participatory rehabilitation and protection of forests,
- Promotion of conservation oriented farming systems
- Capacity building and institutional strengthening of the concerned institutions.

The objectives show a shift in the thinking and approach of project planners. The forestry components included buffer zone planting in degraded areas around demarcated government forests and reserves with mixed fast growing species, rather than monocultures. Farmers or village groups are to be awarded contracts with limited rights for harvesting of wood and other forest products. Besides, it also supports the creation of small timber farms on government lands, providing more extended user rights. On paper, the forest authorities claim only 20% of the sales value of any tree crop and the projects pay 80% of the labour cost involved in establishment of the timber farms. However, there was limited direct interaction with communities while planning and thus they adopted a fairly top down approach. The communities were not involved in prioritising activities or discussing alternative solutions to the land degradation problem. The preliminary project review has found that the resource creation part of the project is satisfactory. But, there is lack of proper policy evolution in support of the people's participation. The participation of farmers is only though payment for work done. The project also lacks a gender strategy, and specific targeting of poor families is limited. Participatory rural appraisals implemented prior to initiating the project activities are of a rather summary nature, and the project offers a largely pre-determined package defined in the project document. This is typical for many projects oriented towards physical targets. A process oriented approach defining the participation and institutional development objectives at the core of the project would probably have focussed differently. This is not to say that it is wrong to have physical targets for rehabilitation of watersheds with critical importance for water supply, electricity and food security of the country, but the strong focus on reaching quantitative targets is seen in this project to have consequences for quality and sustainability. Participation through wage labour does not secure the development of local social and institutional mechanisms to sustain the benefits of the project in the long run.

Many public institutions in Sri Lanka suffer from resource constraints, lack of trained personnel and old fashioned centralised decision-making structures. The Forest Department is not an exception in this regard. The tradition – inherited from the colonial period – has been one of attempting to limit the access to the forest resources in order to secure their protection. Controlling the use of forest resources and enforcing all relevant rules and regulations has not been possible with the resources available. More than half the time of some field personnel is used in court prosecuting forest offences. This way of prioritising limited resources has been encouraged by the system allowing the Department to keep a substantial part of the fines collected for their internal use, and the officer in question can personally retain a similar amount as an incentive.

Rural communities in Sri Lanka often lack strong local organisations, and joint community action related to natural resource management is limited. This makes it natural to focus on family units when developing e.g. timber farms. Plans to enhance community forestry through the UWMP have therefore not gained much momentum. There is lack of proper policy and institutional support for encouraging active participation of the communities.

NATIONAL POLICIES AND SOCIAL FORESTRY

The Forest Policy prescriptions during colonial and neo-colonial periods in South Asia have been influenced by the mainstream views on deforestation leading to restrictions on local communities to extract biomass from state forests in this region (Gadgil and Guha, 1995).

Most of these policies continued in the post-colonial period, but under a different nomenclature. International policy documents on environment and development such as WCED (1987) have also been echoing such views and providing support to national policies in developing countries. These restrictive forest policies and legislation have rather weakened the State in its efforts to conserve natural resources.

In general, forest legislation is south Asia is based on a 'command and control' system, stating mainly what is not allowed, rather than how the forests *can* be used.

The system creates numerous opportunities for corruption, including private sale of produce from the state forest, payments for overlooking offences and unofficial 'fees' for issuing permits for logging and transport. Particularly the latter two have created widespread resentment among local people, and conflicts with the forest agencies. State failures have led to the realisation by the governments and conservationists in South Asia that the orientation and direction of past state policy on forest management had itself been the primary reason for the worsening state of forests. For example, The Indian Forest Policy, 1988 (GOI, 1988) mentioned that:

" Excessive concern for revenue, a disregard for people's needs, and lack of initiative in involving people in the past needs to be corrected ".

Such policy statements need to be highlighted to show that conflicts actually originate in the dissolution of local institutions and ignoring local people's needs.

Local environmental movements in south Asia have also become more influential in the present decade, especially the grass-roots environmentalism based on the legacy of issue-focused local voluntary bodies (e.g., *chipko* movement in India), and the political ecology movement, which presents a wider critique of modernization. Grassroots environmentalism and political ecology have both been able to gain momentum by involving society at the local level. Informal grassroots forest protection movements in south Asia such as the "Western Ghats campaign" and "Save the Himalayas", promoted environmental concerns as well as local interests through media and rehabilitation programmes (Gadgil 1993; Gadgil & Guha 1995).

The pressure from various movements across the region and international agencies eventually led to new forest policies in south Asia during the 90s. The new forest policies of Sri Lanka, India and Nepal appear to be moving towards participatory use and management of forest resources and the involvement of local people, community groups and the private sector. Also, local organisations like the FECOFUN in Nepal, have gained power though the development of a national non-government organisation.

This organisation has been able to influence political decisions, and is putting pressure on forest authorities to allow local management and local retention of revenues. Pressure is also being asserted to reduce corruption within the forest departments. Thus, the push for change in policies and institutions is building up from beneath, and supplements a push for change coming from international cooperation and changes in development policies.

This new approach to forest management attempts to promote people's participation through co-management. This approach could be a part of the new people-centred paradigm recognizing that local people are capable of managing their environment. The new approaches offer a promising opportunity to respond to the numerous problems which states are facing (Poffenberger & Singh 1996) while implementing forestry projects.

However, the approaches are also co-opted by the state and, given the scepticism in the administration to give power to the people, the implementation of the new policies and the participatory role of local people the new paradigms is not clear. Is it simply to co-opt rural people into state-dominated modes of resource use in order to lessen state obligations toward resource management and enforcement, or will local communities really be involved in decision-making?

According to some, the new forestry management paradigms are genuine attempts to involve local people in natural resource management (Sarin 1993; Mukherjee 1995). There are others who are sceptical about it, however, since it has been framed externally in a top-down manner (Kolavalli 1995; Menon 1995).

The top-down nature is evident from the policy documents where the state defines the rules and terms of partnership in forest development projects. The local communities are meant to follow these terms, which do not show much flexibility. For example, the ratio of benefit sharing between forest protection committees and Forest Departments in general are fixed by the forest agencies. Activities such as tree felling and fodder extraction are possible only with the prior approval of these agencies. This also applies to plantations created under social forestry projects on forest as well as community lands.

The major constraint in implementing the new forest policies lies in the hesitation of the State bureaucracy to transfer powers to local communities. This is evident from the presence of the local forest officer who represents the state in the village committees constituted under new forest management models. The local forest officer will not only be a technical advisor, but will also control permits for use and sanctions for misuse of natural resources. The state has rights of exclusion here, since it decides who may or may not have access to forest resources.

The village committees do not have any powers to frame rules of withdrawal or impose sanctions on offenders. The most common approach to resolve offences so far was through paying bribes. Thus the system of permits controlled by state officials was vulnerable to corruption and manipulation, and its continued existence only served to highlight the vulnerability of the state authority. The Forest Service personnel see the new decentralized policy as a threat to their authority. Many problems associated with the new decentralized approach stem from the policy guidelines, which are interpreted to suit the interests of the Forest Service

According to Kothari *et.al.*, (1995), externally initiated community forest management efforts in South Asia have not proved very effective in controlling deforestation problems. With recent socio-economic changes in the region the present situation with reference to biomass extraction may not remain the same in the future if resources are not managed properly. Markets may start influencing forest management in the region encouraging people to cut trees. It all will depend on how the new policy initiatives will actively empower local people and to what extent the people will be involved in management of natural resources.

State initiatives to help local communities in meeting their energy demands through village woodlots or closure schemes have remained as isolated attempts. In a few areas, state forests and village lands that were closed periodically under rehabilitation programs by Forest Departments were found to have good vegetative cover. This may be due to additional protection measures which the department provides to closures, but in general such measures were not popular since they often excluded local people, as seen from the damage to plantations, frequent grazing and low success rate.

CONCLUDING REMARKS

What do we learn from this overview? The linkage between biomass extraction by local people and deforestation is not so easy as it has been portrayed by mainstream views. These theories do not capture the local communities-nature interface. The approach in social forestry was too simplistic in the early stages based on mainstream views; to create block plantations in fenced areas for producing biomass. Biomass problems are more linked to local tenure rights, access to resources, and social and cultural factors, often ignored by planners and researchers.

Besides, the block plantations created new problems like e.g. reduced access to common lands for poor people and landless and violent conflicts. The social forestry cases from the three countries show that there is a change in the approach towards forestry interventions from the time they started. Simultaneously, there is opposition from certain groups who are reluctant to part with the powers and authority.

However, if the impacts of forestry projects on livelihood and forest management are to be sustained such issues have to be addressed. There is an urgent need to empower the local groups and secure access to resources for the landless and poor. Local initiatives to protect forests are hampered by insecure access to resources, and lack of alternative sources of income for the landless and poor.

Despite the reluctance of forest authorities to give up power, the process has gone beyond the point of no return; people are also not willing to give up the opportunities to harvest the forest resources. This is underscored by the legislation and policies, which support local forest management in south Asia, and by the fact that forest users are getting organised with the support of NGOs. Backlashes and delays in the implementation will therefore not change the overall picture, unless the governments are committed to address the issue and actively involve local people. Local initiatives to manage forest resources have to be intensified and integrated in mainstream development of forest management.

2

Ethnomedicinal Plants in Arunachal Pradesh : Some Tacit Prospects

INTRODUCTION

The natural flora and fauna used by the specific ethnic group in the form of medicines in curing and preventing different aliments and diseases are known as ethno-medicine. From time immemorial, indigenous people of different parts of the world have been gathering knowledge about these forms of medicine and medication, which has been handed down for generations merely on verbal tradition. The art is not so compact today because of the allopathic form of modern medicine, which is neither cost-effective nor having easy accessibility at remote Siang villages of Arunachal Pradesh rather still governing the *Adi* way of medication to its people.

The changing aspects of socio-geographic and economic scenario claim a rigorous multi-dimensional study on the problem of starting the farming of ethno-botanical knowledge and plants on ground of their holistic needs of economic, environmental and ethical types. Certainly, the studied area concerned is the world's richest store house of medicinal plants and the people who determine their physique form the totality of nature, still have immense belief on the Mother Nature to a large extent.

It is not the economic botany rather more of pure economics as associated with the tribal medicine; it is at a very young stage of study as far as the studied area concerned. Economic farming of ethno-medicinal plants has been started in the developed nations since a century back and a crude amount of investment coming from both the government and large Pharmaceuticals Companies and Drug Houses is increasing at every minutes of day (Khanuja et al: 2004). Today, the realization comes to our national level planners too from the evidences revealed in the different agro-climatic regions of the world. Instead of having huge potential of farming of medicinal plants at ours own, India is presently importing nearly 90% of total demand from the countries like Nepal, Indonesia, Bulgaria and others. The 10th Five Year Plan (2002-2007) has set a goal to achieve a sum of Rupees 4200 crore targeted economy from the cultivation of medicinal plants. And accordingly, Medicinal Plants Boards at the Centre and State level have been constituted under the Ministry of Health and Family Welfare of the Union Government and State Governments respectively, in order to monitor and to provide the overall guidance for the farming of medicinal plants.

STUDY AREA AND METHODOLOGY

The area of this study is Siang region, named after the mighty Siang river, the down stream of the Tsang-po and upper tributary of the Bhrammaputra, comprising of West Siang, East Siang and Upper Siang districts of Arunachal Pradesh located within the geographical ambit of about $27^0 32'$ N to 29^0 $20'$ N latitudinal and 93^0 $48'$ E to 95^0 $36'$ E longitudinal extents bordering the historic Mc Mohon line to its north and Demaji district of Assam state to its south. The east and west are fringed by the sister districts Dibang valley and lower and upper Sabansiri, respectively. The topographical feature varies form southern plains to the snow clad mountains in the north. The area is located in the subtropical temperate zone with forest coverage of about 63% of the present territorial occupancy, inhabited by 224470 people at a rate of density of 12.12 people per km^2 to its total geographical area of 18518 km^2. The indigenous people is known as the *Adi*, a Mongoloid racial community originated with the Tibeto-Burmese linguist genealogy (Pandey and Tripathy:1997), leads the livelihood basically on forest resources hitherto. Apart form hunting, gathering, fishing and other related activities associated with the forest resources, the *Adis* use to practice *Adiavik*, known as the shifting cultivation or *Jhum* in the Indio logy. Shifting cultivation as a system of cropping practice represents the indigenous from of agriculture being the most commonly and customarily accepted in the region both by virtue of its physical compass and by virtue of heavy reliance of the settlers for their total livelihood. But the practice, which involves cutting and slashing and burning of invaluable forest resources that is hitherto providing a culturally rich bio-diversity of the region, may leads to a severity by courting the extinction of different species of flora and fauna not only having economic values; also it is an alarming rate of subtraction of environmental qualities.

From a general observation and a causal form of oral investigation from the elderly people of the area arising the hope and scope of farming of ethno medicinal plants by which the local farmers may improve their standard of living not only in terms of income generation in the short run, but the practice as proved in the other familiar agro-climatic zones of the world that it could evolve a healthy environment for the mother earth by minimizing the social cost as incurred for *jhumming* and bringing a larger benefit for the society as a whole.

SCOPE OF STUDY

The earliest treatise on Indian Medicine, the *Charak Samhita* (1000 B C) recorded some 340 drugs of vegetable origin and developed the *Indian Materia Medica*. However, since the early 20th Century, rapid extensions of Allopathic system of medication throughout the world generated a huge commercial demand for the pharmacopocial drugs and their products (ICAR: 1992). The actual history of agronomic practice of growing medicinal plants starts form this time. The *Indian Pharmacopoeia* (1966) recognizes some odd eighty-five drug plants grow abundantly and naturally with the prevailing agro-climatic condition in the selected parts of the country. However, no full phased literature on medicinal plants on the studied area is available. As far as Siang region, the land is still a virgin one to the limit of present study. The first ever task of textualization of available medicinal plants in the area was recorded by the Botanical Survey of India in its years long effort. However, the task is yet to be finished because of the difficult terrain and wilderness of the area. Very recently Mibang and Chowdhari: 2003, have recorded the proceedings of the Seminar on Ethno Medicinal Plants of Arunachal Pradesh, where some scholarly works are found. Apart form these, Singh et. al: 2003, and Dhiman: 2003 are also some helpful documentation on medicinal plants of tribal use in general and the studied area in particular. Riba: 2003, has pointed out some relevant socio-economic changes of the tribal life resulted due to a number of reasons, are very much relevant to the context of the present study. However, literature on ethno-medicine is scared as scared for ethno-medicinal plants and economic use of these plants is yet to be revealed in the said region.

Although the ideology of the present study is at the very hypothetical stage, farming of medicinal plants used by the tribal for curing his diseases and ailments is having much economic importance as far as the studied area concerned. The practical evidence of economic farming is yet to be started. The Central Institute of Medicinal and Aromatic Plants (CIMAP) have been carrying on a number of researcher works but Siang region is undiscovered. Very recently, the *Adis* have undertaken the cultivation of medicinal plants successfully in Gagin hills, Lissing hills, Riga-top of Boleng Circle of East Siang District. The District Horticulture Department, East Siang District at Pasighat has undertaken some pilot projects of farming of medicinal plants. It is also providing guidance to the aspiring farmers for introducing the cultivation of medicinal plants, on which they inherited a vast traditional knowledge. An NGO at Mirbuk village at Pasighat has introduced the cultivation of medicinal plants on experimental basis. But private entrepreneurial initiatives of large scale production of medicinal plants, which grow naturally in the study area, are to be found lethargic.

Finding out the economic value of ethno-medicinal plants is a holistic approach indeed and use of those plants as industrial inputs except cinchona and some others is undiscovered to the people. Arunachal Pradesh as a whole and Siang region in particular is literarily having no industry at all. Although the factor endowment and the existing infrastructure for some industrial setup might set another story, industrial development is still at a slumbering state so it is for any Pharmaceutical Plant and Drug House.

The existing scenario of industrial backwardness retards the tremendous growth prospects of medicinal plants at its commercial use. No doubt, the advanced world realizes the healing touch of ethno-botanical plants to a great extent because this form of medicine is proved to be more cost effective and humane friendly up to the extent of some diseases like Aids and Cancer. Plants having medicinal values used by the Aids are not cultivated. These plants grow naturally with no weeding and no effort at all. They are harvested as per household need at any particular point of time. These plants are not so scared so far. A basic hypothesis may gavels that the economic farming is also feasible for these plants in the area by these people.

DISCUSSION

The Arunachal economy as a whole is still succumbing to the severe backwardness for a number of factors, which is not to mention within this limited jargon, but it is not very inappropriate to note that the people are backward because they are backward. For Arunachal economy, agriculture is the prime source livelihood and *jhumming* shares the lion part of the total agrarian practice. *Jhum is* not only the form of cultivation but also a way of life for the Arunachalees. Arunachal Pradesh, being a land locked state of Indian Union with her difficult topographical features does not have a vast scope of setting up large industries and with low level of technological base along with other constraints does not provide avenues for rapid expansion of service sector, may gets a boost if the prevailing agricultural resources are used in a planned way. The Government initiatives in these regard is noteworthy that after achieving the statehood status in 1987, the Government has put emphasis on various types of agricultural extension projects and program like animal husbandry, horticulture, agro-forestry and other allied schemes either as the joint venture with the Union Government or with its own limited resources. There is a number of projects has been being implemented as the substitute of jhum, a primitive form of agrarian practice, but they are not proved to be very effective ones from the ground of its physical relevance. Therefore, the jhumming is going on as it was there at the hundreds of years back. The Jhumias, on the other hand, on the way of carrying out the practice are incurring a huge amount of social cost in terms of different types of natural hazards vis-à-vis, soil erosion, flood and others alike, which can't be compensated in the near future (Riba: 2003). In the jhum land, the Jhumias were once producing merely the amount to sustain rather indulging by the modern life style and other exogenous factors alike, they reduced the spun of jhum cycle in order to produce marketable cash crops. Thus, they are courting more and more ruins for their own. It is evidenced from the present study that the cultivation of medicinal plants as an alternative to jhumming is viable for Arunachal Pradesh as a whole and the studied area, Siang region in particular, where medicinal plants grow abundantly and the ethnic group associated, the Adis, could be better off if they do under take the project, initially with the support of the government and other NGOs, which are working for the upliftment of the society. However, the benefit associated with the farming of medicinal plants in the studied area may be summarized as follows:

1. Medicinal Plants-*An august source of employment generation:* The overall employment scenario in Arunachal Pradesh is as worse as for any other Indian state. Higher growth of population, fast increase of literacy rate along with massive in-migration are the main reasons for increasing figure of unemployment scenario in the state. Since the government with its limited resources is virtually incapable to provide rooms for much employment opportunity in the organized sectors, farming of medicinal plants alike of other agricultural extension programs may open up the scope for unemployed youth to earn their bread from an august source.
2. Medicinal Plants-*An approach to commercialization and modernization of agriculture*: The tribal economy in general is succumbing to its subsistence level hitherto and so it is the Arunachal economy as well. The Jhumias are still producing for their own consumption and no marketable surplus. Although the womenfolk brings a very trivial quantity at the market for raising cash in order to meet the non-food demand, it does not consist a considerable amount of the state Gross Domestic Product (SGDP). On the contrary, if the medicinal plants, which grow naturally in the state could accumulate a fair deal of cash inflow into the economy if they were brought into commercial cultivation applying the modern technology and technological know-how. Thus, the farming of medicinal

plants may be an approach to transform the subsistence Arunachal agriculture into a commercialized one. Another way, it is a stir of modernizing the primitive economy as well.

3. Medicinal Plants-*A stimulant for raising the standard of living*: Growing population pressure along with under utilization of natural resources force the mass to live below poverty line. The farming of medicinal plants may compensate the 69% of the total population in the state, who depends on agriculture and more particularly on jhumming, very often succumbs to crop failure due to different sort of natural calamities. The cultivation of medicinal plans would give them readymade cash to purchase their livelihood. Moreover, it is unlike the household need that they get from the jhum land, the farming of medicinal plants, which will attach the people with market mechanism, would give them better understanding of the outer world.
4. Medicinal Plants-*A catalyst for Industrialization*: Perhaps the difficult terrains and low level of infrastructures hinder the process of industrialization at this north-east extremity of India. Medicinal plants on the other hand, an endowed factor along with traditional knowledge of the *Kennapanna ami* (the village medicine men) may be the economic agents to begin the history of industrialization in the land of the rising sun. The existing Citronella plant at Pasighat in East Siang District may raise the hope further.
5. Medicinal Plants-*The source of Government revenue, not bathos*: A recent study shows that medicinal plant worth of millions of US $ are imported in the country from Nepal annually (Olsen, C.S. et al: 2005). In the era of economic globalization, the phenomenon is virtually contradictory for Indian economy, when we have vast resources for the same products at our own. If the resources are explored properly, import substitution in this arena could save the worthy foreign exchange reserves of millions of dollar. Above all, setting up of industries of medicinal plants will raise huge revenue for the government as well.
6. Medicinal Plants-*Manifestation of indigenous knowledge system*: Primitive local tribal people have vast and extended

knowledge about the various medicinal plants available in this region which certainly helps in conserving the biodiversity.

7. Medicinal Plants-*Protecting the patent rights of indigenous plant species*: Participation in the rat race of globalize economic yard might bring some positive results for Indian economy; rather we loose the rights on *Basmati* or *Neem*, the species of our own. Hence, it is a sort of responsibility to protect the patent rights of the invaluable medicinal plants growing exclusively on the Siang soil.
8. Medicinal Plants-*An ecological agent of immense value! Save our only earth*: The jhum practice, no doubt, feeds the Arunachalees although at subsistence level, but the method of practicing jhum is too cruel to the Mother Nature because it involves slashing and burning of the forest resources without realizing the far reaching consequences of the environment (Sachchidananda: 1989). An alarming rate of deforestation leads to severity for the natural vegetation not only for the area concerned, but the frequent flood and other natural hazards alike mostly caused by this massive deforestation program claim lives and property of billions of rupees as a yearly fiesta. '*Modus operandi* of jhumming first leads to ecological disequilibrium by depleting forest cover, aggravating chances of soil erosion to cause removal of top soil, silting river beds and threatening the existence of life itself and all these by putting grain crop as the only levy on the earth' (Sukla: 1982). However, we cannot escape from the reality that jhumming remains here not merely as a farming method but constitutes as *modus vivendi* of people who operate it. Thus, various alternative projects even did not yield any expected results for the region. Cultivation of medicinal plants, if undertaken commercially, would not only save the natural vegetation but also it is to prevent from flood and other natural disasters caused by the massive deforestation program.

CONCLUSION

The most urgent need for the Arunachalees is to improve the economic condition for the society as a whole. Arunachal, on the other hand, is having vast unexplored resources, which could be utilized in the on going globalize economic condition so amply. The factor endowment of ethno-medicinal plants as industrial input could bring about a drastic economic change of the Arunachal economy, which is not properly realized till the very yesterday. The governmental role in this context is no doubt immense, but the most important is to *raise the entrepreneurial initiatives with development of proper infrastructure* to utilize this scope.

3

Indigenous Medicinal Practices of Bhotia Tribal Community on Indian Central Himalaya

Indigenous knowledge, understood as the unique local knowledge existing within and developed around the specific conditions of people indigenous to a particular geographical area, is gaining more and more attention after having often been rejected as hindrance to development. Indigenous knowledge is almost an essential condition for sustainable development. In India, where the tribes and their tracts constitute very significant parts of the underdeveloped people and area, comprising about 8.13% and 18.70% of the population, and area of the country, respectively. The indigenous knowledge system plays a pivotal role in their very survival. The tribal population in India is about 1.2 times more than the total population of the United Kingdom or France, 2.5 times that of Canada, 4 times that of Sri Lanka, 2/3 that of Bangladesh and more than half as many the population of Pakistan. The tribal communities represent a vast diversity in socio-economic life, cultural heritage and resource use pattern. Despite their habitation in different zones, the tribal people display commonalities in the economic and social life, with variations necessitated to maintain harmonious coordination between the resource availability and population structure.

The Indian Himalayan Region (IHR) represents nearly 18.5% of the total tribal population of India. More than 175 of total 573 scheduled tribes of India inhabit the Indian Himalayan Region. The Indian Central Himalayan (ICH) region known for its dominant Hindu culture; 3.54% of the total population of the region are scheduled tribes belonging to 5 scheduled tribes, i.e. the *Jaunsaries*, the *Tharus*, the *Bhotias*, the *Buxas* and the *Rajis* or *Van-Rawat*, which is a primitive tribe. The *Bhotias*, a transhumant community of Mongoloid origin, inhabit the high altitude regions of the Indian Central Himalaya at Indo-Tibetan and Indo-Nepal borders, a zone of ethnic intermixing and cultural assimilation. They show close racial and cultural affinity to the Tibetans and probably for this similarity the *Bhotia* region is called as *Bod* or *Bhot*, which is synonym for Tibet.

Etymologically, the word *Bhotia* is believed to have originated from the term *bhot* or more correctly *bod*, which means Tibet. The eight major *Bhotia* groups are the *Johari, Jeethora, Darmi, Chaudansi, Byansi, Marchha, Tolcha* and *Jad*, and are scattered over eight main river valleys known as Johar, Darma, Byans, Chaudans (Pithoragarh district of Uttarakhand), Mana, Niti (Chamoli district of Uttarakhand), Nilang and Jadung (Uttarakashi district of Uttarakhand). Each of the sub groups is further divided in to several clans and lineages, which regulate marriage alliances and indicate ancestry. Though, the cultural traits of *Bhotias* reflect close links with the Tibetans acquired through generations of association through trade, they stand distinct from Tibetans with regards to their character and mode of economy.

The culture reveals much closer socio-economic ties with the population inhabiting the middle and lesser Himalayan region of Kumaon and Garhwal. Families are of both joint and nuclear types and both the males and females have equal right on the affairs and decision making system of the family. Traditionally, the *Bhotias* are traders and their trade with Tibet has been the mainstay of their economy. The three main pursuits traditionally followed by the *Bhotias* are the trade, pastoralism and agriculture. The economic organization of the *Bhotias* is an agglomeration of trade and business, terraced cultivation, pastoralism, woolen industry and service. They have two settlements, the upper or summer settlement where they stay from May-June to October-November and cultivate limited varieties of crops like buckwheat (*Fagopyrum esculentum* and *F. tataricum*) and lower or winter settlement where they stay for rest of the year. In lower settlement, they cultivate wheat, paddy, maize, *jowar*, potato, etc. Their herds consist mostly of pack animals like sheep, goat, yak, mules/ponies, etc.

The qualitative relationship of the *Bhotias* with its immediate environs and natural resources has evolved through strenuous experiences of difficult survival. This experiences, being governed by ecology and culture, help evolve tools, technologies and practices for sustenance of the production systems of this community in balance with its social heritage, economic conditions and ecological specificities.

These eco-culturally evolved ecosystem specific tools, technologies and practices in the form of indigenous knowledge constitute integral parts of appropriate innovative strategies that effectively conserve resources and allow options for their optimal use also. Among the *Bhotia* community, the indigenous knowledge governs almost all important productive resource sectors such as agriculture, forestry and animal husbandry.

It revolves around their traditional values of resource use that include subsistence values (food, clothing, housing, medicine, energy), socio-cultural values (ritual, spiritual, aesthetic, educational, psychological), economic-commercial values (agricultural, industrial, pharmaceutical, tourism), and traditional practices of resource use (agri-diversity, wild edibles, medicinal plants, ethnomedicine, forest, grasslands, ethnoveterinary, etc.). The indigenous knowledge of the tribal community serves as a cultural and natural capital, who has a historical continuity of living in harmony with nature with mutual dependence on primary natural resources, and possess a sound knowledge base of the behaviour of the complex ecological system.

The role of indigenous knowledge is vital in the sustainable living of the *Bhotia* community, in view of the fact that replication of modern technologies developed elsewhere has not been successful in the terrain where the tribe live in because of mountain specificities, viz. inaccessibility, fragility, marginality, diversity (heterogeneity), niche (natural suitability) and adaptability. The study aimed at documenting the indigenous knowledge system of medicine and its role in physical well being of the *Bhotia* community, in their resource conservation and socio-economic development. The study tries to explore how the indigenous medicinal system is being able to serve the local people, how it contributes to their scarce economy and how it helps conserve bioresources in around the community.

Methodology

Two sample villages, Darkot and Seepu, inhabited by the *Bhotia* tribal community and located in Munsiary and Dharchula Development Blocks, respectively, in Pithoragarh district of Indian Central Himalaya, synonymous with the state of Uttarakhand, were studied. Village Seepu is a high altitude village being located on an altitude of about 3,586 m amsl, while Darkot is located on an altitude of about 1,237 m amsl. The data on indigenous medicine were collected through interviewing the tribal people using open-ended interviews and guided dialogue techniques. As many as 105 respondents were interviewed independently in their villages to document the prevalent human diseases, their diagnostic knowledge for curing the diseases, and medicinal plants and other raw material used in the treatments. The local names of the diseases and plants with medicinal value were recorded from the respondents while interviewing. Later, the plants with medicinal value were identified with the help of the respondents and taxonomists.

Results and discussion

Evolving over a long period of time based on necessities and experiences, indigenous medicinal system is an important component of indigenous knowledge of the *Bhotia* tribal community, which is an important natural resource that facilitates the development process in cost effective, participatory and sustainable ways and plays an important role in resource conservation. In the studied villages, more than 50 indigenous medicines/treatments are being practiced by the *Bhotia* people in healing more than 45 diseases/disorders using about 40 plant species of medicinal value. While gathering the medicinal plants from their habitat, the *Bhotia* people avoid collecting plants those are infected by insects, pests, and any other disease. Plants affected by toxicity, sunstroke, hailstorms, high velocity winds, fire and floods are also not collected to be used for preparation of indigenous medicines/formulations.

The indigenous medicine and healthcare practices are threatened as the bioresources on which they are dependent are depleting with weakening and even disappearing of institutions and practices of safeguards those have evolved from the cultural contours of the *Bhotia* tribal community to ensure regulated use of bioresources. Further, the ruthless exploitation through unscientific and non-regulated collection of medicinal plants in the recent times by pharmaceutical industries also disturbed the regulatory practices, thereby, threatening and endangering plants of medicinal value. The *Bhotias* follow a number of other regulations, such as maturity of the plants, height of the plants, patterns of branching, colour and other morphological characters, while collecting medicinal plants so as to ensure that the medicinal plants do not die out or disappear from their natural habitat. These regulations are getting diluted under commercialization. Commercial collection of medicinal plants also ignores other regulatory guidelines relating collection of parts of medicinal plants.

For example, the branches are collected when they are fully grown during springs and rains, and young leaves are collected during flowering and ripening of fruits by the *Bhotia* people. Ineffective state management control over the sale of medicinal plants has resulted in over extraction. It is because of the reason that, though, there are prohibitory regulations and restrictions introduced by the Government in the use of medicinal plants and many a medicinal plants are banned for use, clandestine collection of such banned medicinal plants are in regular practice. Seventeen Himalayan medicinal plants, including more than half of them used by this tribal community, are listed in the Red Data Book of Indian Plants.

In addition, several other taxa, although not listed, have reached a stage of critically endangered category. What is required is a secured control of Government over collection and sale of medicinal plants through stringent regulation. The indigenous knowledge system for its own continuation, demands the protection and conservation of bioresources, which is amply reflected in the recommendations of the Convention on Biological Diversity. A successful programme with conservation in focus must keep the above views of Convention on Biological Diversity in forefront and incorporate scientific input compatible with social needs and aspirations embracing societal and cultural principles as base for its foundation. A faithful documentation of indigenous knowledge system and possible value addition will help to the confidence building of the practioners, promote their economy and help to the process of conservation.

Conclusion

This analysis of the indigenous medicinal practices of the *Bhotia* tribal community establishes that indigenous knowledge is a hard-earned experience of the tribal/indigenous community, which ensures physical well being to them, promotes their economy and conserves their resources. Since, the existence of the indigenous medicine is largely dependent on bioresources, the tribes have evolved socially approved regulatory practices and adaptive strategies to conserve the bioresources. However, these regulatory practices and adaptive strategies, that were vibrant, are weakening as the tribal culture is getting assimilated into the dominant global culture of cash oriented society based on market forces and supported bv the Government agencies in the name of development. Simultaneously, there is a need for effective state intervention over use and conservation of bioresources, particularly medicinal plants. This process necessitates assessment of adequacy of laws, policies and action plans promulgated in the above context. A more refined approach could be empowering the tribal people to ensure a culturally sensitive response that will serve to protect the tribe's traditional linkages with their natural resources.

INDIGENOUS MEDICINAL PLANTS AND TRIBALS

The Asian Development Bank (ADB) regional technical assistance project (RETA 5953) on Capacity Building for Indigenous Peoples/ Ethnic Minority Issues and Poverty Reduction has two objectives: to strengthen ADB's Policy on Indigenous Peoples through poverty assessment of these peoples in selected developing member countries (DMCs) and to develop, using participatory methods, an appropriate agenda for action to ensure poverty reduction for indigenous groups.

The specific objectives are to

- conduct a poverty assessment of indigenous peoples/ethnic minorities and examine the DMCs' relevant national policies and legislation, programs, projects, and initiatives pertaining to indigenous groups/ethnic minorities as these address the multifaceted dimensions of poverty;

- evaluate and assess the impact of ADB's interventions that address the vulnerability and poverty of indigenous peoples/ethnic minorities in the DMCs; and
- provide capacity building for government, nongovernment organizations (NGOs), and indigenous peoples/ethnic minority groups so that they can actively participate in formulating action plans as well as in ADB-sponsored programs and projects.

In the Philippines, the task was undertaken by a two-member team, one of whom was assigned within the Cordillera area in the northern part of the country with emphasis on a specific ADB project for indigenous peoples, the Cordillera Highland Agricultural Resource Management (CHARM) project. The other covered the southern island of Mindanao in addition to working closely with the focal agency or the relevant government entity responding to the needs of indigenous peoples.

The guiding forces directing the work of the team were creating an atmosphere of trust and transparency; nurturing interpersonal relations as a ground for constructive cooperation; and adhering to a nonextractive engagement with stakeholders. This approach is consistent with ADB policy the "recognition of the right of indigenous peoples to direct the course of their own development and change."

The two-pronged task of the team consisted of working closely with the Office of the Presidential Advisor for Indigenous Peoples' Affairs (OPAIPA), as specified in the third objective above, and the conduct of a poverty assessment in the three ethnographic regions of Mindanao and in the Cordillera region in Northern Luzon.

Participation in OPAIPA activities was through membership of one team member in a technical working group specifically dealing with the Mechanics and Criteria for the Selection of Commissioners for the National Commission on Indigenous Peoples (NCIP) and the Consultative Body. Activities consisted of a National Consultation of Indigenous Peoples Leaders (18-21 April 2001), Ethnographic Consultations Regional Convenors' Workshop (5-6 May 2001), two ethnographic region consultations (Central Mindanao, 23-24 May 2001; Northern and Western

Mindanao, 29-31 May 2001), and planning and assessment meetings during April-August 2001.

One of the authors was also invited to sit in the Selection Committee for the "long-listing" of applicants for Commissioner positions in June 2001 and to serve as cofacilitator for the OPAIPA Workshop with the New NCIP Commissioners (22-25 August 2001).

The outputs mainly provided support to the OPAIPA. These consisted of the documentation of the two Mindanao consultations, the translation/ consolidation of identified needs of the three ethnographic regions in the area, and the translation/ documentation of some quick-response reports. Inputs were also provided in the design of forms, assessment of activities, deliberations on representation of indigenous peoples, long-listing for the selection of Commissioners, and sharing in the conceptualization of and participation in the final workshop with the new Commission.

Stakeholder consultations and networking with various institutions in Mindanao facilitated the conduct of the Mindanao Indigenous Peoples Consultation on Poverty Reduction and Development on 21-22 July 2001 in Cagayan de Oro City. Results of the gathering provided a broader context to brief case studies of two areas in Bukidnon and Davao, respectively.

The other author was primarily responsible for the evaluation of the policy framework for indigenous peoples' development, analyzing national poverty trends, and investigating the ADB-funded CHARM project. The project covers Abra, Benguet, and Mountain Province. It aims to reduce poverty in the Cordillera Administrative Region (CAR) "by increasing disposable incomes of smallholder farm families in the target areas." In May and June, 2001, stakeholder consultations with indigenous leaders, representatives in the respective communities, and local authorities were conducted. These field consultations resulted in an in-depth analysis of poverty in the Cordillera and insights on the CHARM project. A workshop in the Cordillera was held on 4-5 June 2001.

Information gathered through the participatory approach was corroborated by empirical studies by academic institutions, NGOs, and government organizations. After the fieldwork and other consultations, draft poverty assessment reports were presented and validated during the provincial and national workshops.

The national workshop was conducted on 1– 2 October 2001 at ADB Headquarters, Manila, entitled National Workshop on Mainstreaming Government Programs for Indigenous Peoples' Poverty Reduction. The workshop focused on the dialogue between various government line agencies and NCIP. It also benefited from the constructive inputs of other stakeholders including NGOs, indigenous peoples' representatives, and funding agencies. The plan of action produced by the national workshop corresponds to the draft NCIP framework of operations.

The data presented in this report consist of contributions (workshop consultations, key informant interviews, focus-group discussions) from the indigenous peoples, the OPAIPA, NCIP, NGOs, indigenous peoples' organizations (IPOs), academe, and other agencies. Data from these stakeholders were also made available upon request. Field notes were also shared by anthropologists Daylinda Banzon Cabanilla (on the Ata Manobo), Aileen Toohey (on the Bajao).

4

Defining Indigenous Peoples

DEFINING THE IDENTITY OF INDIGENOUS PEOPLES

ADB, in its Policy on Indigenous Peoples, notes the key concepts of self-identification; linguistic identity; distinct social, cultural, economic, and political systems; and unique ties to ancestral territories. Two significant characteristics of indigenous peoples are observed:

- descent from population groups present in a given area, most often before modern states or territories were created and before modern borders were defined; and
- maintenance of cultural identities; and social, economic, cultural, and political institutions separate from mainstream or dominant societies and cultures. In some cases, over recent centuries, tribal groups or cultural minorities have migrated into areas to which they are not indigenous, but have established a presence and continue to maintain a definite and separate social and cultural identity and related social institutions. In such cases, the second identifying characteristic would carry a greater weight.

ADB uses the following working definition of indigenous peoples in its operations.

Indigenous peoples should be regarded as those with a social or cultural identity distinct from the dominant or mainstream society, which makes them vulnerable to being disadvantaged in the processes of development.

To date, the Philippines is the only country in Asia that has officially used the term *indigenous peoples* and recognized their rights as such. The Indigenous Peoples Rights Act (IPRA), enacted in 1997, defines indigenous peoples as follows.

A group of people or homogeneous societies identified by self-ascription and ascription by others, who have continuously lived as organized community on communally bounded and defined territory, and who have, under claims of ownership since time immemorial, occupied, possessed and utilized such territories, sharing common bonds of language, customs, traditions and other distinctive cultural traits, or who have, through resistance to political, social and cultural inroads of colonization, non-indigenous religions and cultures, become historically differentiated from the majority of Filipinos. ICCs/IPs [indigenous cultural communities/indigenous peoples] *shall likewise include peoples who are regarded as indigenous on account of their descent from the populations which inhabited the country, at the time of conquest or colonization, or at the time of inroads of non-indigenous religions and cultures, or the establishment of present state boundaries, who retain some or all of their own social, economic, cultural and political institutions, but who may have been displaced from their traditional domains or who may have resettled outside their ancestral domains.* IPRA, Chapter II, Section 3h).

Inherent in this definition are factors such as historical continuity, self-identification, and group membership.

Historical continuity is characterized by occupation of ancestral lands/domains, or at least part of them; common ancestry with the original occupants of these lands; culture in general or specific manifestations of it; and language and residence, whether in their lands of origin or evicted from it.

The concept of self-identification is inextricably related to group membership, identity, and history. The thread that weaves these factors together is the indigenous peoples' attachment to land and territory.

NEGOTIATING IDENTITIES

Despite efforts to define indigenous peoples, there is still no consensus as to exactly who are the indigenous peoples in the Philippines. This is due to the absence of up-to-date cultural mapping in this country. Another reason is the tenuous way in which indigenous peoples' identities have been constructed by themselves and others (for political, religious, and other reasons).

This problem of "contesting identities" is best illustrated in the case of the indigenous peoples in Mindanao.

It is not easy to label the identities of peoples of the geographic region known as Mindanao, since the criteria for doing so are not fixed. Even supposedly monolithic categories such as "Moro" are not immutable. Political expression of the Moro as an identity has undergone some variations in recent years. There are also other overlapping layers of identities that cut across ethnic and religious boundaries. Thus, the best that can be achieved is an attempt at categorizing peoples based on (mostly) external criteria, rather than self-identification.

The dichotomy between Moro and Lumad is a historical creation—i.e., a result of the process of Islamization. Until now, common origin stories abound among Muslims and Lumads. In their folklore, Muslim and nonMuslim groups claim the same ancestry. For example, the oral literature of the Subanen of Zamboanga refers to four legendary brothers: Tabunaway, who brought forth the Maguindanao; Dumalandan, who sired the Maranao; Mili-irid, who begot the Tiruray; and Gumabon-gabon who was the ancestor of the Subanen.

The Manobo and Maguindanao of the former Cotabato Province recount the story of the brothers Tabunaway and Mamalu, who were their common ancestors. In the Manobos version of the story, they also share the same ancestor as the Ilyanun, the Matigsalug, the Talandig, and the Maranaw. Tiruray legends also speak of Tabunaway and Mamalu as their ancestors.

The introduction of Islam split the peoples of Mindanao into two distinct categories, Moros and Lumads. Those who adopted Islam became the Moros and those who did not became the Lumads, a Visayan term which means "born of the earth." Lumads are regarded as the original inhabitants of Mindanao. Lumad also now refers to the nonMuslim, nonChristian indigenous peoples of Mindanao.

There are at least 13 Islamized ethnolinguistic groups indigenous to Mindanao. They are the Maranaw, Maguindanao, Tausug, Yakan, Samal, Sangil, Molbog, Kalibugan, Kalagan, Palawani, Iranun, Jama Mapun, and the Badjao. The 18 nonMuslim or Lumad groups are the Bagobo of Davao del Sur, South and

North Cotabato, and Sultan Kudarat; Mansaka of Davao del Norte; Mandaya of Davao Oriental; Subanen of the Zamboanga Peninsula; B'laan of Davao del Sur, South Cotabato, North Cotabato, Maguindanao, and Sultan Kudarat; T'boli of South Cotabato; and Tiruray of North Cotabato, Maguindanao and Sultan Kudarat.

The Manobo encompass various tribes that are also considered Lumad because they are found in Mindanao—in Agusan del Sur and Norte, Davao, and Cotabato.

Other Manobo tribes include the Higaonon of Agusan and Misamis Oriental; Bukidnon of Bukidnon and Misamis Oriental; Talaandig of Bukidnon; Matigsalug of Bukidnon and Davao del Sur; Umayamnon of Agusan and Bukidnon; Dibabawon of Agusan and Davao; Banwaon of Agusan and Misamis Oriental; Talaingod of Davao del Norte; Tagakaolo of Davao Sur; Ubo of South Cotabato; Tasaday of South Cotabato; B'lit of South Cotabato; and Mangguangan of Davao del Norte and Sur, and Cotabato.

In Mindanao society, ethnic identity forms the core of all other layers of socially constructed identities. An individual is identified first by his or her ethnic affiliation. Religion is now also a key element in the differentiation and construction of these identities. Through the years, many members of the ethnolinguistic groups have adopted Islam as a way of life. Some converted to Christianity, while others, the Lumads, held on to their indigenous beliefs, practices, and traditions.

The terms *Muslim* and *Moro* have been used interchangeably to refer to those people who have adopted Islam as a religion and a way of life. However, *Muslim* refers to a universal religious identity, while *Moro* denotes a political identity distinct to the Islamized peoples of Mindanao and Sulu. The Spanish colonizers originally used the term for peoples of Mindanao who shared the religion of the Moors who had once colonized Spain. The term *Moro* was used in the same derogatory way as the term *Indio* for Filipinos from Luzon and the Visayas whom they converted to Christianity.

From the late 1960s through the early 1970s, Moro intellectuals reclaimed the term *Moro* in their efforts to "imagine" a distinct Moro identity. Triggered by the Jabida massacre of 1968, among others, Moro intellectuals led by Nur Misuari created and popularized the *Bangsamoro* identity. *Bangsamoro* became an inspiration, a concept that connoted a peoples' shared identity and vision for a new nation. Such claims to nationhood may not be validated by historical evidence, but the term has gained cogency and acceptance. It has inspired among members of Muslim society a subjective feeling of being one. It has also assumed the following characteristics: it is anticolonial and inspired by a tradition of resistance, it is anti-elite, it is distinctly Islamic, and it is rooted in a struggle for justice.

The Moros expressed their consciousness of a nation through a political struggle for self-determination and in 1969 formed the Moro National Liberation Front (MNLF).

Islam is central to the development of Moro identity. It serves to unite the 13 ethnolinguistic groups that form the so-called "Moro Nation." By interfacing religious doctrines and (Moro) nationalist ideas, the goal to establish the Moro Nation became necessarily a goal to preserve Islam and Islamic culture.

In the 1980s, the project to build a *Bangsamoro* began to manifest some cracks. The Moro political struggle was divided by two opposing concepts of Moro Nationalism. On the one hand, MNLF Chairman Nur Misuari, albeit a devout Muslim, advocated a secular brand of leadership. For him, the role of the MNLF was to uphold social justice and to lead the Moros out of their economic and social ruin. Misuari's goal was "modernist," and was grounded on liberal and political thought.

On the other hand, Hashim Salamat posited that Islam is the primary source of legitimacy of their struggle, and matters of faith and political leadership are not separate realms. He further argued that the *Bangsamoro's* only guarantee for success is inspiration and guidance from the Koran and Islamic rules. For him, leaders should be religious scholars who can apply the correct strategy and directions for the struggle, based on the Koranic scriptures. In 1984, Salamat left the MNLF to form another organization, the Moro Islamic Liberation Front (MILF).

Political analyst and journalist Eric Gutierrezsuccinctly analyzed the cause of the factionalism, thus.

At the root of the MNLF-MILF debates are their differing notions of Moro identity. The modernist MNLF adheres to an inclusive definition of that identity (including non-Muslim Christians and Lumads); while the MILF, on the other hand, adhere to a definition that proposes greater exclusivity for the Muslims. Although it tends to accommodate groups whose existence cannot be denied, thereby confirming the increasing diversification of community life and individual experiences, the MILF strives for moral consensus, internal cohesion and standardization of fundamental beliefs and values within the framework of Islam.

Where do indigenous peoples or Lumads stand amid these acrimonious debates? As may be gleaned in the above quote, both the MNLF and MILF do recognize the existence of Lumads. However, because of the claim for nationhood and the need to rally all *Bangsamoro* to a single national identity, the MILF and MNLF have not been very active in highlighting ethnic identities, to the point of dismissing the issue as irrelevant at this point.

Neither the MNLF nor MILF has clearly articulated a program for asset redistribution, including the indigenous peoples' demand with regard to their ancestral lands. This issue is potentially and actually an explosive one, especially in Mindanao. In the recent peace negotiations, the MILF has listed the recognition of ancestral domain as a primary item in their agenda.

Meanwhile, the indigenous peoples of Mindanao have become more self-conscious and increasingly vigilant of their rights. On 17-19 January 2001, some 67 Lumad men and women leaders from 20 ethnic groups held a "Mindanao Indigenous Peoples Peace Forum" in Davao City. The gathering produced an Indigenous Peoples Peace Agenda, which they addressed to the government's Armed Forces of the Philippines, the MILF, and the New People's Army (a communist militant group). The statement included the following demands.

a. That the Lumad population be sufficiently represented in any peace negotiation in Mindanao, especially in negotiations with the MILF and the NDF (National Democratic Front). The Government, however, should give special support to local initiatives on confiict resolution and transformation, like the local declaration of Lumad territories as Peace Zones.

b. That the Senate and the House of Representatives should pass a law declaring the territories of the Lumads as autonomous. This should include the Lumad territories within the Autonomous Region in Muslim Mindanao (ARMM).

c. That all concerned should recognize and respect the territories and boundaries established by the elders during the D'yandi and Pakang times. This includes territorial agreements among Lumad tribes, between Lumads and the Moros, and between Lumads and Christian settlers.

d. That all concerned should respect our beliefs and practices and our ways of modifying them in the face of Islamic and Christian doctrines and practices, as well as the impact of globalization.

e. That the Government should protect Lumad territories from the incursion of mining, plantations, and other projects that threaten the tribes and their environment.

f. That local government units should pass ordinances, and even release additional funds, if possible, to reinforce the provisions and full implementation of the IPRA.

g. That the liberating forces of the IPRA against discrimination be reinforced through joint efforts of schools, churches, the media, and other institutions.

The whole concept of *Bangsamoro* may also be debated in the light of demographic changes in Mindanao in recent years. As of 1990, there were more nonMuslim (81% of the total population) than Moro occupants (19%) of the Moro terrirory, which is generally considered to include the 14 provinces and 9 cities covered by the 1976 Tripoli Agreement and the 1996 Peace Agreement. Due to past government efforts to move settlers into the region, it is not surprising to find Ilonggo, Cebuano, Tagalog, or Ilocano speakers in one neighborhood with a mixture of Maranaw, Tausog, Maguindanao, and Arabic speakers. Small ethnic groupings, like the B'laan, T'boli, and Manobo, have their own distinct languages as well. Only 5 provinces (Sulu, Tawi-Tawi, Basilan, Lanao del Sur, and Maguindanao) and one city (Marawi) remain "Muslim enclaves." These demographic changes present a major political dilemma for a wider definition of "Moro areas."

In the 1990s, the concept of *tri-people* gained wide discussion in civil society. This concept attempts to grapple with the Mindanao's multiethnic, multipeople identity. The *tri-people* refers to three peoples: the majority Filipino Christian settlers, the Lumads, and the Moros. The concept has been used "not only to describe this diversity, but more importantly, in recognition of a historical process of unification." The *tri-people* concept presupposes that the Christians, Moros, and Lumads consist of three distinct peoples. Thus, the Maguindanao, Maranaw, Tausug, and other Muslim tribes are collectively called the Moro people. The Higaonon, Subanen, Tiduray, T'boli, and other nonMuslim tribes are collectively called Lumads. Those among them who have converted to Christianity are called Christians.

The construction of identities of the peoples of Mindanao continues. Certainly, these identities will be invented and reinvented amid demographic changes, political exigencies, and religious dimensions. But one thing seems constant: the ethnic dimension of identity. Ethnicity will persist as a central axis of this evolving identity—even beyond the completion of the so-called *Bangsamoro* project.

5

The Indigenous Peoples of the Philippines

DEMOGRAPHIC PROFILE OF INDIGENOUS PEOPLES

According to the NCIP, the total population of indigenous peoples in the Philippines in 1998 was estimated to be between 12 and 15 million. However, because NCIP estimates are based on an unofficial count, reference should also be made to the latest complete survey of the National Statistics Office conducted in 1995. Comparing the two sources suggests that the indigenous population might even exceed 20% of the national total. Caution should be taken: the Government may have excluded groups or individuals as indigenous because they did not qualify in the technical definition of the term. It is possible that the actual indigenous population is much bigger.

According to the NCIP, the majority (61%) of the indigenous peoples are in Mindanao while a third (33%) reside in Luzon. The remainder (6%) are scattered among the Visayan islands.

The Summer Institute of Linguistics estimated that there are 171 different languages in the Philippines. Of these, 168 are living languages and 3 are extinct. The same numbers also represent the different cultural entities that speak these languages.

The Philippines was divided into seven ethnographic areas as set by IPRA for representation in the NCIP (see map, page viii):

- Region I and CAR
- Region II
- The rest of Luzon (Regions III, IV[part], and V)

- Island groups (Regions IV [part], VI, VII, and VIII)

Northern and Western Mindanao (Regions IX [Zamboanga Peninsula] and X [Northern Mindanao])

Southern and Eastern Mindanao (Regions XI [Davao Region] and XIII [Caraga])

Central Mindanao (Region XII [SOCCSKSARGEN])

Within the ethnographic areas defined above are different indigenous peoples exhibiting varying social, cultural, political, and linguistic features. Each of these indigenous groups has its own history of contact with the dominant Filipino society and this has placed each group somewhere on a spectrum from "isolation" to "acculturation."

In Region 1 and CAR, the indigenous peoples are concentrated in the northern mountain ranges of the Cordillera occupying the interior hills, narrow strips of flat land along deep valleys, and plateaus. The Tingguian, Isneg, and northern Kalinga are found in the watershed areas of the Abulag, Tineg, and Chico rivers. These groups are largely swidden (slash-and-burn) cultivators who depend on farming rice, root crops, and vegetables. Along the slopes of Mount Data and nearby areas are the Bontoc, Sagada, Ifugao, and Southern Kalinga. They are wet-rice cultivators who grow their crops both on terraces and swidden fields. The Ibaloi and Kankanaey inhabit the southern region of the Cordillera. Like their neighbors to the north, their subsistence economy is based on wet and dry agriculture. In more recent years, these groups have been integrated into the outside market economy with the growth of commercial gardening of temperate vegetables.

In Region 2, several lowland indigenous groups like the Ibanag, Itawes, Yogad, and Gaddang inhabit the Cagayan Valley of northeastern Luzon. The Caraballo range at its southern end is home to the Ilongot, Ikalahan, Isinai, and some Aeta groups, which continue to carry out wet-rice agriculture, swidden farming, hunting and gathering, and some commerce.

There are other indigenous groups in Regions III, IV, and V. Along the area of the Sierra Madre range of eastern Luzon are the Dumagat, Pugot, and other Aeta groups whose economies have depended on swidden agriculture, hunting and gathering, fishing, and trade. Communities of Paan Pinatubo (sometimes called

Hambal, Ayta, and Sambal) live under the shadow of the Zambales range, and Baluga inhabit the northwestern part of Pampanga and the southwestern area of Tarlac. Along the Pacific coast in the province of Quezon, Pollilo Island, and the Bicol peninsula are Agta groups, known as Kabihug (Camarines Norte), Agta Tabangnon, Agta Cimarron, and Itom (Camarines Sur, Albay, and Sorsogon). Increasingly, these groups have depended on cultivation, fishing, gathering of forest products, and wage labor.

The island groups include the seven Mangyan groups in the island of Mindoro: the Iraya, Alangan, Tadyawan, Tao-Buid, Buhid, Hanunoo, and Gubatnon. In the interior foothills and remote coastal areas of Panay and Negros Islands are the Sulod and the Ati. In the island of Palawan, the Tagbanua are found in northern and central areas. The Batak are found in small pockets north of Puerto Princesa City and in northeastern Palawan. The Palawanon are in the southern part of Palawan near Brooks Point.

Generally, the Mangyans of Mindoro; the Sulod and Ati of Panay; the Aeta groups in Negros Island; and the Batak, Cuyonen, Palawanon, and Tagbanua of Palawan practice swidden agriculture, combined with hunting, fishing, gathering, and trade.

Along the hinterlands and coastal lowlands of the Zamboanga peninsula live the Subanen, while toward the uplands of northern Mindanao, particularly on the plateaus of Bukidnon, are the indigenous peoples who call themselves Higaonon meaning "mountain dwellers."

Within the highlands of central Mindanao, particularly along the mountainous slopes of the provinces of Davao, Bukidnon, and Cotabato are the Bagobo. Groups that occupy the upper headwaters of the Davao, Tinanan, and Kulaman rivers are the Manuvu, Matigsalug, Ata, and others.

In the coastal areas along Davao Gulf and interior hinterlands of southeastern Mindanao are the Mandaya, Coastal Bagobo, Agusan Manobo, and Ata.

For the most part, the indigenous peoples in Mindanao basically subsist through swidden and wet-rice cultivation, hunting, fishing, gathering, and the trade of locally manufactured items.

Outside influences have caused changes in the economic systems of indigenous peoples. The consequences for the communities involved are varied, and it seems that efficient accommodation of elements from the new system is often facilitated by innovative social organization that ensures control of the land by the indigenous community. For example, when the Igorots of Sagada were introduced into the western market economy, the gradual transformation did not entail much conflict because the Igorots have maintained control over their own resources and become involved in social relations that are not commoditized. Similarly in the case of the Buhid of Mindoro, government economic and developmental interventions have not resulted in the outright assimilation and economic deprivation of the group because of ingenuities in landownership among them.

In other cases, the introduction of a different cultural system led to the disintegration of indigenous society. For instance, the promotion of American rule among the Tiruray of Cotabato transformed them into peasant communities, as immigrants introduced new lifestyles and a market economy. The same could be said of the Higaonon of northern Bukidnon Province, who suffered abuse and exploitation from immigrants. After independence, the Philippine Government established a greater presence in the formerly remote areas of indigenous occupation. This required that communities of indigenous peoples elect village captains, mayors, and national leaders. All of this created changes in the leadership structure of the hitherto isolated communities. However, there remain groups little influenced by change and others where the imposed structure and traditional structure continue to coexist.

Among the Agta, the network of intertwining familial and kinship ties continues to provide assistance, privileges, and protection to its members. Leadership among the Bungkalot, Teduray, and Sulod is assumed by the elder who possesses the personal skill to persuade and influence, and who is a good speaker with knowledge of customary law. The council of leaders, called *intugtukan* among the Bontoc, *tontong* by the Ibaloi, and *pakilukesen* by the Arumanen Manobo is composed of persons of wealth and charisma, and with knowledge of customary law. They meet to settle cases and decide on important social and religious affairs.

The influence, power, and authority of the Datu prevails among the Manuvu, Higaonon, and some Bagobo groups. He is assisted by a council of leaders and other respected members of the community in the performance of his obligations. Among the functions of a Manuvu Datu are settling cases, keeping the peace, and deciding where the village is to move for the next swidden cycle.

STATE POLICIES TOWARD INDIGENOUS PEOPLES: A HISTORICAL OVERVIEW

The indigenous peoples in the Philippines are, to a large extent, a creation of history. Prior to Spanish colonization, indigenous communities already had customary concepts and practices of land use and ownership. The indigenous concept of landownership was one of collectivism, where private ownership by one or a few individuals seemed alien.

The customary laws of the ethnolinguistic groups embodied this basic rule, but there were also variations. Among the Manobo, Mamanua, and Hanunuo groups, land was considered as communally owned by the people forming the community. The Ibaloi concept of ownership was one of sharing the land with the gods, ancestors, kindred, and future descendants. But to the Subanon, only the paramount god, Apo Gumalang, had the right to own land. This was also the case with the Kalingas, for whom only Apo Kabunyan, the high god, could claim ownership over land. These people saw themselves only as caretakers of their resources.

The indigenous peoples' strong attachment to their land and resources led to intermittent conflicts (the so-called tribal wars) with other groups (kin) that violated their territory. At varying levels, these communities had developed social and political structures to regulate their relations within their communities as well as with outsiders. Intercommunity relations ranged from cooperation to conflict. However, the idea of majority-minority dichotomy and its attendant problem of marginalization and discrimination were absent. These problems emerged during the colonial period.

With the advent of Spanish colonization, the "minorization" of the indigenous peoples started. The Spanish colonizers forced their subjects to live in *pueblos* through a policy called *reduction*. Those natives that refused to live in these pueblos retreated into the hinterlands and were called *remontados* and *infieles*. The natives of the mountains, like the Igorots of the Cordillera, put up a strong resistance against colonial intrusions against their territories. Because of this, they were called *tribus indipendientes* by the Spanish chroniclers. They were also labeled as barbarians, pagans, and all sorts of derogatory names. Soon, even the assimilated *indios* internalized these prejudices against indigenous peoples. Thus, a dichotomy between the assimilated majority and unassimilated minority emerged.

The Spaniards introduced laws that contradicted, even denied, customary concepts of land use and ownership. They introduced the Regalian Doctrine, first, through a policy of *encomienda*. These were land grants by the King of Spain, but were managed by an *encomiendero*. The latter was tasked to collect *tribute* (taxes) and enforced the economic policies of the Crown *(bandala, polo y servicio,* etc.). Later, the Spanish Crown enacted the Maura Law, which reiterated that all pueblo lands were protected lands; they could not be alienated because they belonged to the King. In spite of these colonial policies, the unassimilated indigenous peoples continued to practice their customary practices with regard to their land and resources.

The Americans essentially retained the concept of the Regalian doctrine. They made laws that reinforced the state's control over the public domain, citing the reason, among others, that there was no effective system of land registration during the Spanish period. The laws passed during that period included the following.

- Land Registration Act No. 496 of 1902, which declared all lands subject to the Torrens system and empowered the State to issue to any legitimate claimant a proof of title over a parcel of land.
- Philippine Commission Act No. 178 of 1903, which ordered that all unregistered lands become part of the public domain, and that only the State had the authority to classify or exploit the same.
- Mining Law of 1905, which gave the Americans the right to acquire public land for mining purposes.
- Public Land Acts of 1913, 1919, and 1925, through which Mindanao and all other areas of fertile lands that the State considered unoccupied, unreserved, or otherwise

unappropriated public lands became available to homesteaders and corporations, despite the fact that the indigenous peoples were in these lands.

Postcolonial administrations of the Republic of the Philippines did not veer away from the western concepts of land use and ownership. The 1935 Constitution stated that all agricultural, timber, and mineral lands of the public domain, waters, minerals, coal, petroleum, and other mineral oils, all forces of potential energy, and other natural resources of the Philippines belong to the State. This constitutional mandate gave rise to a number of laws that continue to deny indigenous peoples' customary land use and ownership. The most controversial of these laws was Presidential Decree (PD) 705 of the Revised Forestry Code, enacted in May 1975.

PD 705 further delimited landownership among indigenous peoples, so that even ownership of rice terraces of the indigenous peoples of the Cordillera (for example), became questionable. This decree stated that "no lands of the public domain with a slope of 18% or more shall be classified as alienable and disposable, nor any forest land 50% or more in slope, as grazing land. Lands 18% or more in slope that have already been declared as alienable and disposable shall be reverted to the classification of forest lands... to form part of the forest reserve...that when public interest so requires, steps shall be taken to expropriate, cancel effective titles, reject public land applications, or reject occupants thereof." PD 705 made the indigenous peoples of the Cordillera virtually "squatters in their own land," because most of these lands are located in mountainous areas that are within the 18% slope rule.

The 1987 Philippine Constitution also retained the Regalian Doctrine. Section 2, Article XII, of the 1987 Constitution states that all "lands of the public domain, waters, minerals, coal, petroleum, and other mineral oils, all forces of potential energy, fisheries, forests or timber, wildlife, flora and fauna, and other natural resources are owned by the State."

At the same time, the 1987 Constitution also contained provisions that recognized "the rights of indigenous cultural communities within the framework of national unity and development" (Art. II, Sec. 22) and the creation of autonomous regions in Muslim Mindanao and in the Cordillera (Art. X, Secs.

15-19). However, the state's constitutional recognition of indigenous rights "in the context of national development" did not fundamentally change the situation of the indigenous population. This could be seen in the implementation of government policies and programs in their territories, like reforestation programs, the National Integrated Protected Areas System (NIPAS) Act of 1992, Mining Act of 1995, and Official Development Assistance (ODA) projects.

Massive reforestation programs, which are funded by multilateral finance institutions, are linked to commercial tree planting activities. Those taking place in indigenous peoples' lands consider the encroached territories as forestlands, in which the inhabitants are requested to plant certain species of trees. Because of this there is no available useful area for them to continue agricultural activities needed for their own subsistence. Often, people are also displaced when government forces conduct military operations in the area to deal with pockets of resistance to the projects.

The NIPAS Act was to protect plant and animal species that are endangered, or have the possibility to become such. The law aims to establish national parks as designated areas for the preservation of biological resources. Its conservation scheme, which excludes local residents while entrusting the areas to state bodies, has limited access by indigenous peoples to the territories in which they carry out their subsistence activities. The provisions for people's participation in protected areas management seem to be confined to an initial level of consultation and not as far as the identification, planning, and implementation levels.Likewise, it is possible for the new bureaucratic structures that accompany the NIPAS implementation to challenge the indigenous sociopolitical system in the area.

When the Mining Act was signed into law in 1995, it essentially allowed greater control of the Philippines' geological resources by transnational and local mining corporations. The agreements embodied within the law, like the granting of royalty payments for the right to mine in indigenous peoples' lands, have favored the apparent expropriation of these lands as the local people yield to corporate mining interests. Displacements of indigenous peoples have also resulted from mining operations that necessitated the transfer of occupants. Among those affected are the B'laan of North Cotabato, Subanen of Zamboanga del Sur, Aetas of Central Luzon, Igorots of the Cordillera region, and some groups in the Cagayan Valley.Environmental disasters from mining activities, like the Marcopper mine accident in Marinduque Province where mine tailings found their way into rivers, have sparked a series of protests against the Mining Act, and continue today.

ODA projects carried out during the last quarter of the 20th century to improve the economic condition of the people and to promote social justice were also funded by foreign bilateral and multilateral institutions. The projects have resulted in a number of positive outcomes for the expected beneficiaries, including increased land productivity, a rise in net incomes, and improvements in access to water supply. However, there have also been negative outcomes. In the Cordillera, there have been violations of indigenous land rights, restructuring of the local economy to the advantage of outside market forces, and absence of some expected outputs of projects.

In 1997, following the 1987 Constitution's mandate, Republic Act (RA) 8371 of the IPRA was enacted. A landmark piece of legislation, the IPRA promises to change the course of history of indigenous peoples in the Philippines.

6

The Indigenous Peoples Rights Act

In October 1997, then President Fidel V. Ramos signed into law RA 8371, the IPRA. This legislation lays down the legal framework for addressing indigenous peoples' poverty. It seeks to alleviate the plight of the country's "poorest of the poor" by correcting, by legislative fiat, the historical errors that led to systematic dispossession of and discrimination against indigenous peoples. The Implementing Rules and Regulations (IRR) of IPRA were approved on 9 June 1998. The IPRA law enforces the 1987 Constitution's mandate that the State should craft a policy "to recognize and promote the rights of indigenous peoples/ICCs within the framework of national unity and development," and "to protect the rights of indigenous cultural communities to their ancestral lands to ensure their economic, social and cultural well-being." KEY ELEMENTS OF THE INDIGENOUS PEOPLES POLICY

The IPRA provides that development programs, projects, and activities must be developed along the fourfold agenda of recognition and protection of ancestral domain/land rights, self-governance and empowerment, cultural integrity, and social justice and human rights.

Right to Ancestral Land/Domain

The law restores the rights of indigenous peoples over their ancestral lands and ancestral domains. The term *ancestral land* under the IPRA refers to lands occupied by individuals, families, and clans who are members of indigenous cultural communities, including residential lots, rice terraces or paddies, private forests, swidden farms, and tree lots. These lands are required to have been "occupied, possessed, and utilized" by them or their ancestors "since time immemorial, continuously to the present." (Section 3 b).

Ancestral domains are defined as areas generally belonging to indigenous cultural communities, including ancestral lands, forests, pasture, residential and agricultural lands, hunting grounds, worship areas, and lands no longer occupied exclusively by indigenous cultural communities but to which they had traditional access, particularly the home ranges of indigenous cultural communities who are still nomadic or shifting cultivators. Ancestral domains also include inland waters and coastal areas and natural resources therein. Again, these are required to have been "held under a claim of ownership, occupied or possessed by ICCs/IPs, by themselves or through their ancestors, communally or individually since time immemorial, continuously to the present." (Section 3 1).

The law enumerates the rights of indigenous peoples over their ancestral domains/lands. These are the right of ownership over the ancestral lands/ domain, right to develop and manage lands and natural resources, right to stay in territories and not to be displaced therefrom, right to regulate entry of migrants and other entities, right to safe and clean water, right to claim parts of reservations, and right to resolve conflicts according to customary law (Sections 1-8, Rule III, Part II).

Indigenous peoples also have the right to transfer (ancestral) land or property among members of their community, and to redeem those that have been acquired from them through fraudulent transactions (Sections 1, 2, Part III).

The law also stipulates that the indigenous peoples and their communities have the responsibility to maintain ecological balance, restore denuded areas, and "adhere to the spirit and intent of the Act."

Right to Self-Governance and Empowerment

The IPRA upholds the right of indigenous peoples to self-governance and empowerment. The IRR further define the range of the exercise of these rights. These are the right to pursue their economic, social, and cultural development; to use commonly accepted justice systems, conflict resolution mechanism, peace building process, and customary laws; to participate in decision making; to maintain and develop indigenous political structures; to have mandatory representation in policymaking bodies; to determine their own priorities for development; to establish their tribal *barangay* (village) and equivalent voting procedures; and to organize people's organizations.

Indigenous peoples may now also utilize a vital instrument for their empowerment—the principle of free and prior informed consent (FPIC). Section 3, Part III of the IPRA states that

The [indigenous peoples] *shall, within their communities, determine for themselves policies, development programs, projects and plans to meet their identified priority needs and concerns. The [indigenous peoples] shall have the right to accept or reject a certain development intervention in their particular communities.*

The indigenous peoples' decision to accept or reject a proposed policy, program, or plan shall be assessed in accordance with their development framework and their value systems for the protection of land and resources.

Right to Cultural Integrity

The IPRA and its IRR put forward procedures and mechanisms for the recognition of indigenous peoples' right to cultural integrity. They incorporate such provisions as the constitutional and legal framework for the right to cultural integrity, the conceptual framework for cultural integrity, the specific rights pertinent thereto, and the procedures for the recognition of these rights, including the right to indigenous culture, customs, and traditions; right to establish and control educational learning systems; recognition of cultural diversity; right to name, identity, and history; community intellectual rights; protection of indigenous sacred places; right to protection of indigenous knowledge; and the right to science and technology.

Social Justice and Human Rights

Recognition of and respect for fundamental human rights are also safeguarded by the IPRA. The law contains specific provisions that ensure that indigenous peoples, just like other human beings, will enjoy these rights: the right to life, development, and civil liberties; political rights; freedom of association; nondiscrimination; equal protection; and right to peace and social justice.

The law guarantees indigenous peoples' right to basic social services. The indigenous peoples have the right to employment, vocational training, housing, sanitation, health, social security, infrastructure, transportation, and communication.

THE NATIONAL COMMISSION ON INDIGENOUS PEOPLES

The IPRA, following a constitutional mandate (Art. 16, Sec. 12), created the NCIP as "the primary government agency for the formulation and implementation of policies, plans and programs to promote and protect the rights and well-being of the ICCs/IPs and their ancestral domains as well as their rights thereto."

The IRR spell out the procedures for the organization of the NCIP, including its creation, agency, composition, appointments, qualifications, tenure and compensation of personnel, appointment of Commissioners, powers and functions, and ethics and codes of conduct of its officers and various offices.

The IRR also define procedures to institutionalize indigenous peoples' participation in the NCIP through a Consultative Body as provided in Sec. 50 of the IPRA. A placement committee will oversee the merger of the Office of Southern Cultural Communities and the Office of Northern Cultural Communities in the NCIP organizational structure.

Implementation of the Indigenous Peoples Rights Act

Four years after the enactment of the IPRA, stakeholder assessment of its implementation yields dismal results. The NCIP, the government body that is mandated to implement the law, has been unable to perform its tasks well. Participants at the Orientation-Planning Workshop for the New NCIP Commissioners, held on 22-25 August 2001 at Tagaytay City, assessed the current status of NCIP as follows.

- Lack of clear leadership has led to the absence or nonperformance of NCIP's policymaking and adjudication functions as well as coordinated delivery of basic services.

- NCIP's present structure and staffing have hampered the institution from carrying out its functions.
- NCIP officials are by and large not qualified or equipped with the necessary skills for NCIP to meet its mandate under the IPRA.
- There were indications that the processes of issuing titles and certificates to allow mining and other activities have been compromised if not corrupted.
- Past policies and decisions have been detrimental to the operations of NCIP and these should be identified and changed to better serve indigenous peoples.

A concrete manifestation of NCIP's weakness has been the slow pace of processing and approval of Certificate of Ancestral Domain Title (CADT)/Certificate of Ancestral Land Title (CALT) applications. As of February 2001, the NCIP national office had approved only 9 of the 181 CADT applications, and 347 CALT applications. The Tagaytay workshop findings validate numerous reports of alleged irregularities in the issuance of CALTs. These reports claimed that several CADT/ CALT applications were haphazardly approved, and did not go through the procedures mandated by the law.

In September 1998, a Presidential Task Force on Ancestral Domains (PTFAD) was established. The PTFAD was assigned to study the issues related to the composition, administrative set-up, and operations of NCIP. The memorandum also directed the Department of Budget and Management to withhold the release of NCIP's operational funds. These reduced the capacity of NCIP to process and approve ancestral domain titles.

After submitting its recommendations to the President, the PTFAD was dissolved. However, the Government created two new bodies, the National Anti-Poverty Commission on Indigenous Peoples and the Presidential Task Force on Indigenous Peoples. The former was instructed to address poverty-related issues affecting indigenous peoples while the latter was an advisory body on policy matters pertaining to indigenous peoples. These functions obviously overlapped with those of the NCIP.

In June 1998, NCIP produced the IRR. One of the most important and controversial provisions of the IRR is Section 5 (6),

Rule II which states that "the ICCs/IPs shall have the right to suspend or stop any project or activity that is shown to have violated the process of securing free and prior informed consent, or have violated the terms and conditions of such previously granted consent."

The mining Industry immediately expressed their apprehension over this NCIP rule. They also questioned the IRR's bias for indigenous peoples and their priority claims over the country's minerals and other natural resources to the detriment of other sectors like the mining and power industries. The Chamber of Mines of the Philippines indicated that there had been an alarming exodus and withdrawal of foreign mining companies due to the "inherent vagueness of the IPRA" and the apparent "lack of commitment by the Government to actively support the development of the mining industry." The mining industry even threatened to bring their issues against the IPRA to the Supreme Court.

NCIP issued an Administrative Order (No. 3, 13 October 1998) exempting all leases, licenses, contracts, and other forms of concession within ancestral domains existing prior to the promulgation of the IRR, from the coverage of IPRA's provisions on free and prior informed consent. That Administrative Order also declared that all written agreements with and/or resolutions by indigenous peoples' communities prior to the IRR shall be considered as "free and prior informed consent." NCIP may also issue a temporary clearance to individuals and companies pending resolution of conflicts within ancestral domain areas.

Advocates and indigenous peoples' organizations censured the new Administrative Order for being "onerous" and a "sell-out" to the interests of the mining industry. They posited that "property rights" as defined, while still being debated, seem to be interpreted in favor of the mining interests; that the short duration (30 + 7 days) allowed to secure FPIC is insufficient; that allowing "interim clearances" and any form of written agreements to pass off as consent is fraudulent; and that the idea of doling out "gifts" in the form of showcase development projects before and during the conduct of FPIC is divisive and deceptive.

The greatest challenge against the IPRA came on 28 September 1998, when retired Supreme Court Justice Isagani A. Cruz, together with Atty. Cesar Europa, filed a petition before the Supreme Court. The petition contained three main points: first, that the law violates the constitutional principle that all natural resources belong to the State; second, that the law deprives the State of control over the exploration and development of natural resources; and third, that the law threatens to strip private owners of their properties.

After a thorough deliberation, the Supreme Court voted on 6 December 2000 to dismiss the petition. However, the separate decisions of the Supreme Court justices who voted to dismiss the petition expressed substantive points that essentially upheld the Regalian doctrine. They stressed that the IPRA merely gave indigenous peoples surface rights over their ancestral domains. The resources therein, they add, are still owned by the State.

On 20 February 2001, newly installed President Gloria Macapagal-Arroyo issued Executive Order No. 1, creating the OPAIPA, delineating its functions to ensure the effective implementation of the IPRA. The OPAIPA was assigned to review the performance, qualifications, and fitness of NCIP officers and staff and to institute a process for the proper selection of staff of NCIP, as mandated by RA 8371. Following nationwide consultations, the President appointed a new set of Commissioners in August 2001.

The appointment of the new Commissioners through a consultative process and the full support of several indigenous peoples' organizations, civil society, and President Macapagal-Arroyo has rekindled hopes for the successful implementation of the IPRA.

Impact of the Indigenous Peoples Rights Act

Four years after its enactment, the IPRA has mixed effects on indigenous communities. What follows is an attempt to highlight the advantages as well as the disadvantages of the IPRA on indigenous communities.

The long years of legislative action leading to the IPRA's approval and the ensuing debate during its implementation have undoubtedly increased awareness of the public (including the indigenous peoples themselves) on the issues and concerns of indigenous peoples. They have drawn the attention of the State as well as civil society (nongovernment organizations (NGOs), churches, and academic institutions) to the "plight" of indigenous peoples, who had long been denied a space in national/mainstream Philippine politics.

Several new indigenous peoples' organizations (IPOs) and national federations and alliances emerged during the period. At the national level, organizations/ alliances were established to give support to the IPRA. One of them was the *Katutubong Samahan ng Pilipinas,* which was formed through the initiative of the Philippine Association for Intercultural Development (PAFID) and the Legal Resource Center. Another broad alliance was the Coalition for Indigenous Peoples Rights and Ancestral Domains (CIPRAD). This is a network of 15 IPOs and five NGOs. Two of these NGOs are the lawyers' group known as *Tanggapang Panligal May sa Katutubo* (PANLIPI) and the Episcopal Commission on Indigenous Peoples of the Catholic Bishops Conference of the Philippines. The Manila Office of the International Labour Organization (ILO) played an important role in the establishment of CIPRAD. Lending "critical support" to the IPRA is the National Coalition of Indigenous Peoples of the Philippines. This was spearheaded by the Davao-based Lumad Mindanaw and supported by the *Sentro para sa Ganap na Pamayanan.*

At the community level, IPOs were organized and community solidarity enhanced as a result of community efforts to delineate ancestral domain and ancestral lands. A case of a "good practice" was the community-based Bakun Indigenous Tribes Organization (BITO) in Bakun, Benguet, supported by the ILO-Inter-Regional Programme to Support Self-Reliance of Indigenous and Tribal Communities through Cooperatives and Self-Help Organizations (INDISCO). BITO was officially accredited to lead in managing the ancestral domain of the Kankanaey and Bago peoples in Bakun who were awarded a Certificate of Ancestral Domain Claim in 1998. While they wait for the final approval of their CADT, the organization has come up with development activities like creating sustainable livelihoods and community-based businesses that do not harm or threaten Bakun's ancestral domain.

While the IPRA has resulted in the proliferation of IPOs and support NGOs, it has also engendered disunities among the indigenous peoples. One source of discord is how to view the IPRA. The IPOs and NGOs that support the IPRA argue that, despite its imperfections, they are convinced that this legal instrument can be used as a stepping-stone toward a "more progressive level of political discourse." The Cordillera Peoples' Alliance does not share this view. They labeled IPRA a "master act of deception." They decried the persistence of Regalian doctrine in the law.

The negative effects of the IPRA are felt most at the community level. Since IPRA was promulgated, there have been many community-level disputes. In the Cordillera, for example, there has been an increase in boundary disputes over the past four years, attributed to longstanding conflicts over resources (e.g., water for irrigation) and territorial delineation. The IPRA exacerbated these conflicts.

In the province of Abra, members of a clan who belong to the Masadiit tribe are fighting over the areas that should be covered by their respective "ancestral domains." One faction of the clan now resides in the municipality of Sallapadan, while the other faction lives in the municipality of Bucloc.

Because IPRA stipulates that indigenous peoples should resort to customary laws to resolve conflicts, the two factions are "recreating" and retelling their respective version of their customary law. People of Bucloc have accused the Council of Elders *(lallakay)* of Sallapadan of being senile; thus, their genealogical reckoning and rendering of oral history should not be taken seriously.

The IPRA has also resulted in the "construction" or "imagination" of ethnic identities. In Romblon, for instance, the Bantoanons of Banton Island are now claiming that they are indigenous peoples. They assert that they should enjoy the rights enshrined in the IPRA. Some sectors have questioned this, saying that the Bantoanons are, in many ways, not different (not distinct) from the Batanguenos.

There is also the problem of ancestral domains being "imagined." This is a result of a generalizing concept of ancestral domain, that is, the notion that ancestral domain is a static or fixed concept of a communal territory, with persistence of indigenous sociopolitical institutions. In reality, many indigenous peoples have already adopted and adapted western property regimes and do not care about domains.Some indigenous peoples have opted for Torrens titles and members of some clans are fighting each other for private titles.

Still, some experts view the IPRA as an instrument that will lead toward privatization of the "commons."For example, in Agusan del Sur in Mindanao, a certain Datu applied for some 76,000 hectares of ancestral land.Indigenous peoples expressed fear that privatization will facilitate the entry of corporations and business persons who will just negotiate with one individual to be able to enter the indigenous communities.

These different views suggest that while the framework for addressing indigenous peoples' concerns is already established by the IPRA, much needs to be done. For one, there is still a need to put in place monitoring and regulatory mechanisms to ensure that commercial interests do not unduly appropriate indigenous peoples' lands. There is also a need to adopt an area-specific and culture-specific ancestral land/ domain policy that takes into consideration local nuances, processes, and tenurial systems.

7

National Dimensions of Indigenous Peoples' Poverty

An important issue to resolve is whether indigenous peoples in the Philippines are disproportionately poor. In other words, is the incidence of material poverty for indigenous communities more serious than for other sectors of the national population? It would also be useful to determine poverty trends over a period of time. A crucial issue, of relevance for the formulation of future poverty reduction measures, is whether there are arguments for targeting programs specifically at indigenous populations.

On this important point, owing to existing data limitations, the evidence can only be anecdotal. It is obvious that indigenous communities that have been relocated from their traditional communities and that have lost access to their land, natural resources, and other assets, are likely to bear the burden of material poverty as well as sociocultural deprivation. The same is true for communities who have lost their livelihood as a consequence of mining, logging, and similar extractive activities.

As regards the relationship between ethnicity and poverty however, there are no reliable data available at the national level. In the absence of this, the authors' method has been to extrapolate from national aggregated data from the National Statistics Office. The preliminary finding, using this method, is that poverty is not exclusive to the regions with populations that are predominantly indigenous peoples. In 1997, for instance, while the average incomes of the National Capital Region (NCR) and Central and Southern Luzon are indeed disproportionately high in comparison with all other regions, Caraga, the poorest of the indigenous peoples' regions, performed as badly as Eastern Visayas and Bicol regions. CAR and Regions X and XI performed better than all of the regions in the Visayas, using average income as the indicator. This is corroborated by qualitative data from certain parts of Mindanao where indigenous peoples insist that they are not poor. Their main concerns are better presented as the growing risk of impoverishment, as their traditional lands and natural resources come under increasing threat.

On the whole, in the indigenous peoples' regions, the incidence of poor families and poor populations did not improve substantially between 1988 and 1997 despite the rise in average income, except for Region II. In some instances, the incidence has worsened and, ironically, in the indigenous peoples' regions that registered the more respectable growth rates in average income. This implies that the fruits of regional development have not "trickled down" to the poorest of the poor. Moreover, in the indigenous peoples' regions that registered a high incidence of poverty, the poverty gap ratio, especially in the rural areas, is also stubbornly high. These two measures show the extent of income inequality. The target annual poverty thresholds have to be raised in order to ensure that more families and persons are assured of annual per capita incomes that satisfy basic nutritional requirements. There do appear to be cases of significant growth rates for some indigenous communities, but the existence of marked inequalities within these same areas impedes fair access to the fruits of economic growth.

In the next section some welfare indices are examined to locate the situation of indigenous peoples' communities vis-à-vis the national situation in order to explore where policy intervention may be most effective. The sections thereafter flesh out the statistics that are used in the indices by presenting actual cases of the poor contending with low incomes, high prices, and scarce social services.

WELFARE INDICES

Indigenous peoples' regions, for the purposes of this report, were taken as those in which the indigenous population is at least 40% of the total regional population. Table 1 shows that there are five such regions in the country.

INDIGENOUS PERSPECTIVES ON POVERTY AND DEVELOPMENT

It is important to observe that Philippine indigenous notions of *poor* and *poverty* generally differ from western "modernist" concepts of these categories. Most indigenous peoples do not want to be labeled as poor mainly because of the negative and discriminatory connotations of this construct. Mainstream characterization of the poor includes images of sloth and ignorance, which are used by "development" agencies and corporations to justify exploitation of the land and resources of indigenous peoples.

For indigenous peoples, development is not a matter of accumulating material wealth. On the contrary, they have cultural mechanisms (called by anthropologists "redistributive mechanisms") to prevent a member of the community from accumulating too much wealth.

Of course, these indigenous notions of poverty and development have changed through time. With the advent of "modernization" brought about by missions and formal education; construction of roads; the entry of migrants; and the introduction of television, transistor radios, and newspapers (as well as other print media), new values have emerged. However, indigenous values persist or are modified to adapt to the changing conditions. Indigenous peoples' views of poverty manifest the dynamics of continuity and change.

In the following two chapters, the characteristics of poverty as viewed by indigenous peoples in two regions, the Cordillera and Mindanao, and assessments of its extent in those regions are presented.

POVERTY ASSESSMENT

This section attempts to convey the perspectives of the poor in the Cordillera with regard to poverty and development. It also aims to present a general picture of the poverty situation in the region. This is a result of 4 months of field work in the form of focus-group discussions and key informant interviews in the provinces of Abra, Benguet, and Mountain Province. The field data are augmented by studies conducted by academic organizations and NGOs.

First, the general poverty situation of the Cordillera is presented. Next are perspectives of indigenous peoples of the Cordillera on poverty, followed by views on the causes of poverty. The fourth section tackles the indigenous notions of well-being and their expressed needs and aspirations. The final section outlines the perceived causes of poverty.

GENERAL POVERTY SITUATION

The Cordillera region of northern Philippines contains the highest and largest mass of mountains in the country. It is home to over one million indigenous peoples collectively known as Igorots. It has five provinces: Benguet, Mountain Province, Ifugao, Kalinga, Apayao, and the city of Baguio. The indigenous peoples by major ethnolinguistic identity are Ibaloy, Kankanaey, Bontoc, Ifugao, Kalinga, Isneg (Apayao), Tinggian, and Kalanguya. The region is rich in natural resources: rivers and hotsprings; minerals, mainly gold and copper; lowland and temperate agricultural areas; some remaining forests; and high biodiversity.

The native inhabitants of the Cordillera are indigenous peoples who have occupied and sustained their territories since time immemorial. They have maintained their culture, distinct languages, and sociopolitical systems. Not having been effectively colonized by Spain spells the difference between a Filipino minority like the peoples of the Cordillera and the majority of the colonized Filipinos. This difference persisted even with American colonization and succeeding Filipino regimes that pervaded all aspects of socioeconomic and political life in the Cordillera. The indigenous institutions persist even as they accommodate external influences.

Another feature of the Cordillera region is that it is one of the poorest and most marginalized regions in the country. All the Cordillera provinces are among the country's poorest provinces, more popularly known as "Club 20."

As of 1997, there were 110,000 families or about 42% of all Cordillera families living below the poverty threshold of P12,836. This means that almost half of all Cordillera families did not have sufficient income to meet their food and nonfood requirements. The incidence of poor families decreased from 51.0% in 1994 to 42.5% in 1997. This was attributed to the 59.9% increase in average family income in the region between 1994 and 1997. Moreover, the total number of poor families decreased by 12,800 or 10.4% over the same period. Despite this improvement, with about 4 out of 10 families living below the poverty line, CAR ranks 5th among the regions with the highest incidence of poor families.

There remains a wide disparity between the poverty situation in urban and rural areas. In 1997, the incidence of poor families in the urban areas was 14%, while that in the rural areas was 55%. This disparity is attributed to the slower growth of average income in the rural areas (32%) compared with urban areas (72%) from 1994 to 1997.

The 1996 Minimum Basic Needs Survey conducted by local government units in CAR revealed that the greatest unmet needs in the Cordillera are income and employment, followed by water and sanitation, then by basic education and literacy.

Health

The total population of the CAR was estimated at 1,404,000 in 2000. The annual growth rate has been highest in Benguet (2.9%), Ifugao (2.3%), and Apayao (2.2%). The birthrate in 2000 was 22.25 live births per 1,000 population. The infant mortality rate in 1999 was 13.82 per 1,000 births, with the main causes being pneumonia, preterm birth, septicemia, congenital anomalies, and respiratory distress syndrome. Most provinces have good immunization schemes. However, malnutrition has been increasing, with 9% of preschool children classified as either moderately or severely underweight in 1999 compared to 5% to that in 1998.

Maternal health care continues to be a problem. Not all pregnant women have been able to avail of prenatal care and vitamin supplements from the health units. Education on pregnancy and childbirth seems to be a continuing necessity to uplift maternal health in CAR.

Potable water continues to be a problem in most provinces of CAR. Virtually all households in Baguio City and Mountain Province have safe water, but only 65% of those in Apayao have access to potable water. Most households in Baguio City (85%) have complete basic sanitation facilities including provisions for sewerage and garbage disposal, but only 19% in Kalinga and 34% in Ifugao have these facilities; the average for CAR is 48%.

For the CAR population, the leading causes of death were pneumonia, cancer, cardiovascular diseases, accidents/trauma, tuberculosis, hypertension, vascular diseases, and peptic ulcers. There is a growing incidence of "affluence-related" diseases.

Among the infectious diseases, the specter of tuberculosis continues to haunt the region. A greater number of cases has been reported from the outlying provinces than in the more urbanized areas. There is more success with leprosy. The Department of Health Integrated Regional Field Office received the Kagawad Award of the Civil Service Commission in 1998 due to its successful leprosy program in CAR.

Malaria, rabies, pneumonia, and gastroenteritis are significant in some provinces of CAR. As of 1999, there were 73 physicians working in the local government units of CAR. There were also 37 dentists and 32 medical technologists, while the numbers of other health personnel (nurses, midwives, and village health workers) can be said to be barely adequate.

Literacy/Education

The Functional Literacy, Education, and Mass Media Survey of 1994 revealed that the Government needed to put more efforts in bringing education to CAR. In terms of simple literacy (i.e., able to read and understand a simple message in any language or dialect), CAR scores poorly, with only 89% literacy rate.

WHO ARE THE POOR?

Each of the ethnolinguistic groups in the Cordillera has its own terminology to refer to the *poor* and *poverty*. For example, the Ibaloi of southern Benguet and the Kankanaey of northern Benguet refer to the poor as *ebiteg* and *nabiteg*, respectively. Both terminologies refer to one who has no land to till and lacks the resources to be able to work. These resources are capital, transportation, machine sprayer, and *kuliglig* (a small tractor). An *ebiteg* is also one who lacks time and money to learn new skills and ideas. He/she is usually indebted, without a regular source of income, and does not enjoy amenities such as television, appliances, or an "LPG" (liquefied petroleum gas for cooking).

The Kankanaey of Mountain Province call the poor person a *kudo*. He/she is one who has no *payew* (irrigated rice field) and, therefore, works on the land of the landed or *kadangyans*. Since the *kudo* is frequently without enough food and money, he/she is forced to stay with well-off or better-off relatives.

The Tinggian of Abra identify the poor as *pobre* or *panglaw*. This refers to a person who is *sangkasapuluan, sangkaapuyan,* which literally means hand-to-mouth existence. They lack food and money to buy basic commodities and send children to school, they cannot read or write, and their children are not properly clothed. These problems occur because they lack such resources as land and draft animals

Kankanaey Traditional Notions of Poverty and Wealth

To understand the Kankanaey notions of poverty, one must understand their notions of wealth or who are considered rich.

Traditionally, those considered rich are those who host thanksgiving feasts called *sida* or *pedit*. These religious feasts are offered to gods and the spirits of the ancestors. These feasts are ranked according to the number of pigs the host family can offer to the gods and spirits. Ranking starts from 3 pigs, then 5, 7, 9, 13, 15, 17 and so on. The sequencing is always based on odd numbers because of the belief that the pig without a pair will again attract more pigs when the host family raises pigs again after the feast.

The bigger the feast a family hosts, the wider the circle of relatives and clans can be invited. The host family gains more prestige as it invites a wider circle from within and outside the community. With the prestige comes the status of *kadangyan* or *baknang*, which connotes wealth and the ability to share such wealth. One was thus considered rich if he/she was able to hold feasts, during which the wealth was shared with many people in the community. One may have dozens of cattle and livestock but cannot gain the community's respect without sharing the wealth through a feast.

Being rich is not synonymous with accumulating wealth. Being rich is being able to share one's wealth through a *sida* or *pedit*. It follows, therefore, that poverty, at least according to old tradition, is the opposite of the Kankanaey notion of wealth: being poor means being unable to invite a wide circle of relatives and clan members to a feast.

Notions of poverty are socioculturally defined. For example, the Tinggian of Abra believe that one indicator of poverty is the physique of the person, i.e., being thin is an indicator of being poor, while being stout is an indicator of being rich. One Tinggian elder even suggested that the smell of a person is a marker of poverty: "You can tell that one is poor when he/she stinks—he/she cannot even afford to buy bath soap."

Current (government) concepts of poverty tend to disaggregate the poor according to income brackets. Thus, the participants of the project's provincial workshop held in the Cordillera identified three categories of poor people. These are (i) ultra poor—those earning below P20,000 per household per annum, (ii) poor—those earning P20,000 to P39,000 per household per annum, and (iii) near poor—those earning P40,000 to P51,000 per household per annum. This categorization is based on the regional poverty threshold of P12,836.

While there may be divergence in definitions of poverty based on the cultural context of each ethnic group, the most common markers are landlessness; lack/no source of cash/income; lack/absence of equipment such as spray machines, *kuliglig* (tractor), and transportation; and lack/absence of access to basic services such as health and education.

CAUSES OF POVERTY

Several studies have shown that most farming communities in the Cordillera (i.e., the interior villages under a subsistence system) experience low production. A 1994 household survey conducted by the Cordillera Studies Center of the University of the Philippines College in Baguio City showed that the average harvest of farming households in 5 communities studied was low. Samoki had the lowest yield per household at 2.5 cavans (first crop) (1 cavan = 50 kilograms) and 2.22 cavans (second crop). The average rice yield was 10.90 cavans per household in Bineng, 31.44 cavans per household in Suyo, and 20.57 cavans (first crop) and 13.33 cavans (second crop) per household in Masikil. The average area of irrigated holdings is 0.5 hectares.

Low production is attributed to several factors, the most important of which are (i) small, fragmented landholdings in most Cordillera villages; (ii) inability of households to generate cash required to purchase inputs; (iii) farmers' limited access to markets because of the poor quality of social infrastructures, e.g., transport and communications; and (iv) inadequacy of social support services and information about these services, such as credit and marketing.

Small Fragmented Landholdings

According to data compiled in 1991, of a total land area of 1,829,368 hectares, 81.4% were public/forest lands, of which 33% (or 26.9% of the total land area) were protected areas and 67% were production areas, of which 6.8% (or 3.7% of total land area) were agricultural areas including public land. Alienable and disposable land constituted 18.6% of total land area.

Land-use data for the Cordillera region reveal that forestland constitutes 40%; extensive land use (cultivated/open areas, grassland, and mixed crops/ shrubs), 53%; and intensive land use (croplands and others), 7%. About 57% of the pine forest area in the Cordillera has a slope greater than 50%, which DENR considers unfit for development (by virtue of PD 705).

These data clearly show that land use extends into what the Government calls " public land." At least some parts of this must include what indigenous communities call their "ancestral lands."

To compensate for the problem of small, fragmentary, agricultural land area, farmers resort to intensification through the use of inorganic inputs or expansion into forested areas. Both options are not without ecological as well as health costs.

Lack of Irrigation

Related to the problem of small and fragmented land is lack of irrigation in most Cordillera farming villages. Data from the National Irrigation Administration show that, despite the fact that there was an increase of 13,000 hectares of land irrigated from 1988 to 1992, overall irrigation service coverage in CAR remained very low at 34%. Among CAR provinces, Mountain Province has the lowest irrigation service coverage (17%) and Abra has the highest (65%). There are still 118,476.66 hectares of arable land that require irrigation to make them productive.

Inability of Households to Generate Cash

The low level of productivity of the upland agricultural systems is also due to the inability of households to generate cash to purchase inputs as well as consumption goods. The 1994 household survey by the Cordillera Studies Center showed that in the 5 communities they studied, farming was the principal source of cash receipts.

Cash receipts of farming households have to meet both production/marketing costs and consumption expenditure. There are several requirements that compete for the cash pool of farming households: debt payment of previous/current cropping period(s); operating capital for the succeeding crop; current household expenditures on food, clothing, housing, education, health, and leisure—immediate needs that cannot be postponed; and savings for fixed agricultural capital investment or for other forms of household investment. In view of these competing uses, and where the cycle for crop production receipts and expenses and the cycle for consumption expenses and receipts do not coincide, farmers are left with two options: off-farm/ nonfarm and other sources of cash receipts, or credit including remittances from absent family members.

Production, whether for home consumption or market, depends not only on land and other natural resources but also on the households' ability to obtain commercial production inputs. On the consumption side, quality of life (i.e., the ability to consume beyond staple goods/services) suffers when the ability of the household to participate in the market for goods and services (e.g., education, health, and recreation) competes with production and basic consumption requirements such as food.

Inadequate Social Infrastructure

A major problem for Cordillera farmers is their limited access to markets because of the poor quality of infrastructure such as transport, storage, and communication. One respondent in Bucloc, Abra, said that because the farmers could not bring their produce to the market, often they just fed it (fruits and some vegetables) to their pigs.

Poor social infrastructure is seen as the "missing service sector" that contributes to the "failure of the market" in most Cordillera farming villages. Market failure occurs when it (the market) cannot allocate resources to production and distribute the output of the process most efficiently. This results in higher equilibrium prices of commodities or factors, lower quantity in the exchange, and decline in overall welfare level:

Here, if you were small farm operator without transport and storage facilities and you have a product with a short shelf life, you are figuratively at the end of a gun pointed at you by a buyer who has those facilities. And the big wholesaler himself, or through a network he has established, is in the best position to exploit the missing market elements.

There is, therefore, a direct link between poor infrastructure and the frequently-lamented problem of high interest rates on credit supplied by entrepreneurs.

Needs and Aspirations of the Poor

I have simple aspirations. I only wish that I had a more regular employment or source of income like a piggery so I will be able to buy basic needs like rice, sugar, coffee, salt, and food. Rosita, 45 years old, from Maguyepyep, Sallapadan, Abra.

We need capital to start a handicraft business here in Bucloc. We have abundant bamboo; all we need are skills trainings and seed capital...We also need capital for livelihood activities like pig and goat raising, and furniture making. Participants of the focus-group discussion held in Lamao, Bucloc, Abra.

"Gawis ay panagbiag": *good life means having my own land to till, being able to buy basic needs, and being able to send my children to school.* Farmer from Fidelisan, Sagada, Mountain Province.

Inadequacy of Credit and Marketing Services

The high cost of inputs and the lack of adequate capital for production are frequently mentioned sources of livelihood instability. The absence of capital limits the number of crops to one per year, or prevents farmers from buying inputs to support regular farm activities. Consequently, income and living standards are negatively affected.

Several studies on the vegetable-growing communities of southern Benguet and Mountain Province (areas that are already producing cash crops at an intensive level) show that a typical household engaged in vegetable production is usually trapped in a cycle of indebtedness because of market problems including low prices, high transport cost, high input cost, and poor access to credit.

The absence of capital should have encouraged rural banks to participate actively in economic activities in the vegetable areas. However, the findings of a 1983 study appear valid today. These studies reveal several problems with the banks: lending policies are rigid and complex for a farmer to observe; rural banks have credit portfolios too small to support both long-term and short-term agricultural production; and banks require credit collateral like land titles, generally uncommon in the Cordillera.

Entrepreneurs and other informal credit sources become ready sources of capital. Unlike rural banks, these informal credit sources have simple lending procedures, share the risk with the farmer-borrower in case of a crop failure due to natural calamities and poor prices, and are willing to lend production credit to farmers twice or thrice. The role of the entrepreneur as buyer, supplier, and creditor explains the high interest rate in the informal sector. Despite this high interest rate, farmers prefer to borrow from entrepreneurs. Thus, because of these informal credit sources, farmers have been able to sustain their production.

INDIGENOUS PERSPECTIVES ON WELL-BEING

The poor people of the Cordillera, like people everywhere, desire a "good life" or at least a better life. The poor's perspectives of well-being are expressed in such terms as *gawis ay biag* and *nasayaat nga biag* (Kankanaey), meaning being able to meet basic needs like food (rice, sugar, salt) and clothing. At the same time, perspectives also vary across ethnic and socioeconomic groups (perhaps even gender).

For some Ibaloy respondents of this study, the Tuba Benguet, good life means the state of being able to buy amenities such as gas for cooking, a tractor to lighten the burden of farm work, and a vehicle to transport their goods and products.

Well-being means good health. Among the Ayangan of Ifugao, good life is synonymous with a healthy life—*ap-ap-hochan,* which means that a person or a community is able to maintain a harmonious relationship with the environment and supernatural world. To attain a good life, one should not ravage the forest or spoil the land, especially the ricefields. One should nurture them (not leave them idle or unproductive), otherwise, one will become ill and suffer misfortune.

For the Iffialig of Barlig, Mountain Province, notions of well-being are also linked to notions of health. The "generic" term for health is *apu-apuor,* which mean absence of illness or sickness. Related to the concept of *apu-apuor* are the terms *kursi/kunog,* which mean strong. Women and men in Barlig have a positive definition of health. Health is not simply the absence of illnesses but a condition of well-being or being strong in order to face the day-to-day challenges of life in a cheerful way.

The Iffialig description of a healthy person is *amma ammay chi achor,* which means that the whole body is in good condition. "When one is conditioned, one is healthy and strong." Note that there is the implicit assertion here that one is able to work or function because the body is fully developed in relation to one's need to work. Health is then considered to be an integrated aspect within the system of life of the people. For many men and women in Barlig, there is emphasis on the idea that "life is work." If one is *apu-apuor, nakursi* or *ammay chi achor,* then he or she is capable of doing things. As one woman emphasized, "one cannot work properly when one is not healthy."

Although indigenous worldviews persist, they do not constrain people in the villages of the Cordillera. Values and ideas of good life are increasingly influenced by what they read, see, and hear outside their villages. Increasingly, people believe that education is a key to a better life, especially for the younger generation. They work hard to be able to send their children to college. Some dream of sending one or two of their children to work overseas.

NEEDS AND ASPIRATIONS OF THE POOR

The following is a "wish list" of needs and aspirations of the poor in the Cordillera, as articulated by respondents as well as by other empirical studies in the region. The list does not cover all the needs and aspirations of all poor people, only those that are most commonly or frequently mentioned.

Education of Children

Studies across the Cordillera region show that the most common aspiration of farming families is a college education for their children, even if they can barely afford the cost. It is common to find farming households selling a piece of land or a precious heirloom to pay for their children's school expenses. Consequently, expenditure for agricultural production is sacrificed. Respondents explain that a college degree for their children is their passport out of farming. This finding suggests that some farmers in the village see the attainment of the "good life" as being outside the confines of their village. Indeed, outmigration is a common phenomenon all over the Cordillera.

Income-generating/Livelihood Activities

This study validates the National Economic and Development Authority (NEDA) survey findings on Minimum Basic Needs: that people in the Cordillera aspire to have more income to provide for basic household needs like food, clothing, education, and medical services. The most common livelihood activities in the "wish list" of farming households are pig raising, fruit and vegetable production, handicrafts business, and dressmaking (for women).

Food Security

Food security seems to be a contentious issue in the Cordillera, according to the views of participants and respondents in this study. Even household surveys are not unanimous on the issue. A 1986 sample survey that looked into food adequacy in the Cordillera (except Abra) revealed that Cordillera households experience shortages in the supply of staple foods or sources of carbohydrates (rice, sweet potatoes, cassava, corn, and taro) and even protein. Some 84% (n = 46) of the respondents claimed shortages in rice and 64%, shortages in sweet potatoes, an immediate substitute for rice. These shortages were not only because of seasonal fluctuations in the availability of food, they constituted a real shortage of food, because all possible substitutes were also in short supply.

A more recent (1994) survey of 5 communities in Benguet and Mountain Province study found that in Bineng (Benguet) and Suyo (Mountain Province) at least, some 85% of the respondents reported that their respective households enjoyed adequate food. On the whole, rice sufficiency (i.e., ability to obtain rice, whether home-produced or bought) was claimed by all households in all study sites, and sufficiency in vegetable by 93%. The study concluded that the problem is not overall food consumption level per se, but insufficiency only in the consumption of particular food crops. For example, insufficiency in consumption of fish/ marine products was claimed by 90% of respondents in Bineng, 77% in Samoki, 72% in Lam-ayan, and 60% in Suyo.

Improved Infrastructure

Participants in the focus-group discussions expressed a common desire for better access roads and bridges, so that they will be motivated to be more productive. The present study's field work confirms earlier studies that reflect the local people's views on the importance of their control of and participation in the running of infrastructure projects in their locality. This means that the local residents' needs, rather than the needs articulated by (sometimes even genuinely concerned) external agencies, should be given priority. It also requires the inclusion of local knowledge, technology, and materials in project planning and implementation, and the subsequent reliance on local expertise to maintain infrastructure.

Participation in the Development Process

Community participation in the development process is a recurrent theme in academic studies as well as workshops and conferences of indigenous peoples' organizations in the Cordillera.

A study conducted in 1993 found that generally, people in the Cordillera do not accept programs, projects, and activities that are "externally determined." This explains their negative reactions toward so-called development projects such as the San Roque hydroelectric dam and several mining interests in the region. Externally determined development projects like these are perceived as having no substantial benefits to the members of the communities, and threaten their ancestral lands and cultural heritage.

Increasingly, people from the communities want more meaningful participation in the development process. Likewise, NGOs and government organizations working at the grassroots level have realized that the participation of community stakeholders in the conceptualization, planning, implementation, and monitoring/evaluation of projects is key to the success of projects.

This study reveals that indigenous peoples in the Cordillera do not give a single definition to the categories "poor" and "poverty." These concepts are given meaning within the context of culture and society. Yet, Cordillera culture(s) and society are transforming, even if indigenous worldviews persist. Poor Cordillera farmers are increasingly exposed to new values and tastes, and even definitions of "good life" are becoming a mix of indigenous and modern values. They do not simply want to be in constant harmony with the environment and supernatural beings; they also express the need for more income to buy such basic commodities as salt and sugar. They also gain more income to be able to afford college education dream of owning a piece of land to till, a carabao or a of their children. A son/daughter who obtains a college tractor to lighten their burden in the field, and a vehicledegree is considered the passport out of poverty. to transport the products. They need all these so as to.

8

Poverty Assessment

GENERAL POVERTY SITUATION

Most of the indigenous peoples in the Philippines are in Mindanao, where 23% of the country's population reside. The island's contribution to total poverty in the country is approximately 31%. Information from various studies reinforces the general assessment about the poverty situation in the island. The provinces where poverty is deepest and most severe contain populations of indigenous peoples. Western Mindanao, which was ranked second nationwide in terms of rural poverty incidence, forms part of NCIP's Northwestern Mindanao ethnographic area. In 1995, the population of the region was about 1.1 million, of which roughly 46% were indigenous peoples.

Preliminary results of the 2000 Family Income and Expenditure Survey show that poverty incidence in Mindanao is higher than the national average of 40%, ranging from 46% in Southern Mindanao to 74% in the ARMM. Poverty incidence was worse in 2000 than in 1997.

According to a recent ADB study based on income measures, the regions with the highest poverty incidence are in Mindanao: Caraga (55%) and the ARMM (57%). Poverty, as revealed in the ADB study, continues to be a rural phenomenon with rural areas contributing 74.1% to total poverty. Western Mindanao ranks second highest in terms of rural poverty incidence. Urban poverty is most severe in Caraga (43%) and the ARMM (51%).

To understand the full dimension of the phenomenon, the ADB study looked into poverty depth and the poverty gap nationwide. In Mindanao, the poverty gap and severity were most severe in Caraga, with five other Mindanao regions (Western Mindanao, ARMM, Northern and Central Mindanao, and Southern Mindanao) also above the national averages.

The 1997 UNDP Philippine Human Development Index Report estimated Human Development Index (HDI) values in Mindanao ranging from 0.41 in the ARMM to 0.62 in Southern Mindanao, all lower than the national average of 0.66. There were however, differences among provinces even within the same region, depending on the type of indicator used. In Southern Mindanao, for instance, Davao del Norte had a functional literacy rate of 85%, while in Davao del Sur it was 69%. All provinces in the ARMM registered very low HDI values. Western Mindanao and the ARMM registered extremely low values in access to safe water, toilet facilities, electricity supply, and health care services.

Potable water is scarce in most of Mindanao, with only 51% of the population in Agusan del Sur and 87% in Agusan del Norte having access to this important commodity. The proportion of families in the provinces of Bukidnon, Misamis Oriental, and Davao del Norte that live in makeshift housing is greater than the national average.

The Social Weather Stations conducted two surveys on self-rated poverty and self-rated food poverty (March and July 2001). In Mindanao, the incidence of self-rated poverty as perceived by residents was 64% in March and 70% in July. Both figures are higher than the national averages of 59% in March and 66% in July. Self-rated food poverty was also higher in Mindanao than the national average in both surveys.

With reference to public services, as noted in the HDI for the area, Mindanao continues to be poor. A recent World Bank study reported the following.

- In health services, 11% of respondents (the highest among the "other island regions") said that the lack of government facilities forced them to use private clinics/hospitals.
- In education, 41% of school-age children do not attend school (compared with the 11% in the NCR, 14% in the Visayas, and 33% in Luzon), citing economic reasons and poor performance.
- With regard to water supply, most households rely on

private wells and rainwater collectors. Only 23% avail of level III sources (considered to be the best water sources).

To aggravate the situation, armed conflict prevails in Mindanao, resulting in the dislocation of more than half a million families to date in Central and Western Mindanao as well as the ARMM. Although the incidence of conflict seems to be isolated and span only a limited time, the impact and reach are felt even in relatively peaceful parts of the island. Business and investments continue to slow down. As stated by Secretary Deles, "The current conflict has been identified as having contributed to the sluggish stock market, falling peso, and weak investor and tourist confidence. And for those in war-torn areas, very few economic activities exist and few of them may prosper, because of the constant occurrence of armed conflict."

THE CONCEPT OF POVERTY AMONG INDIGENOUS PEOPLES

Walay pobre sa una, aduna la'y mga kakulangon. Dili pobre karon, aduna la'y pagkalisodlisod. Nagkalisodlisod kay gilisodlisod. ("In the past, no one was poor. There were only a few unmet needs. Neither are we poor now. We experience some degree of hardship. We live in difficult circumstances brought about by outside forces.")

The quotation above aptly summarizes the perception of Mindanao indigenous peoples' representatives regarding their being labeled poor. At the same time, it acknowledges that with the continuing situation, they will soon be poor.

The indigenous peoples have generic terms that approximate the term *poor*. Some examples identified by specific groups are: B'laan—*tao telando;* Mamanua— *adyu-adyu;* Ubo Manobo—*kaayayuwan, makairo-iro, wara-wara, kahirapon, kaaya-aya-an;* Teduray— *miskinan, enda enda;* Bagobo Tagabawa—*kandaan;* Mandaya—*al'lang;* and T'boli—*el-el.* There is further need, however, to subject these terminologies to more validation in terms of their levels of meanings.

Generally, the term *poor* is considered not to apply to Lumads, but to the *langyaw* (outsider). Viewed as a discriminatory construct, it implies sloth and the ignorance of and use of weapons by outsiders to gain control of resources and lands of the former. The indigenous peoples assert that they are not poor, but are living in difficult circumstances brought about specifically by loss of land and access to resources. They can only rise from this condition if they are able to gain control of their territories again.

Poverty indicators used by outsiders refer mainly to material aspects such as money, house, adequate clothes, and transportation; other aspects include the inability to seek medical attention from hospitals, absence of formal schooling, and the characteristic of being lazy.

When asked whom they consider poor, Lumads refer to squatters who have no land; the Bajao, who are the most oppressed group; streetchildren; prostitutes; underpaid workers; evacuees; and marginalized farmers and fishers. The characteristics of such people are the same as in the outsiders' definition: no property, dirty-looking, clothed in rags, rundown or broken house, children unable to go to school, and having debts to pay.

TERMINOLOGY AND CHARACTERISTICS

The Lumad indicators cover a broad range and mix of indicators, one of which shares the outsiders' perspective of lack or absence of material belongings. Other issues considered important by indigenous peoples for improving quality of life are: powerlessness and dependency, ignorance of indigenous knowledge systems, loss of affinity with ancestral domains, and failure to belong.

Slavery (Mangguangan—*allang;* Mansaka— *gupisan; Subanen—ghulipon)* is considered a state of poverty. Individuals or families have to be provided everything. Consequently the slave can be stepped on and taken anywhere, can be subjected to eviction anytime, and cannot make any independent decision. The individual may even need to be provided a wife by the "master."

Other individuals considered poor are those who are dependent on others. They are unable to earn a living, can eat only if employed, are unable to pay debts, and have to live with others because they have no house. Some of them survive mainly on wild plants and animals as well as root crops. They are timid and need to be protected (Ata Manobo—*ubong-ubong).* If they commit an offense, they are bailed out by others or by the Datu.

Another indication of a state of poverty among the indigenous peoples is loss of concern for the ancestral land and domain. The Higaonon, for example, lament the fact that in the 1970s some of their community members lost land to migrants for a minimal amount or for goods of little value such as cloth, sardines, and alcoholic beverages.

Failure to Belong

Poverty is also not being of royal lineage (Manobo—*raig-raig*) or even being an outsider or not of the group (Manobo—*saruwa*). The poverty stigma also holds true for one who is banished by the group for unacceptable behavior (Subanen—*ginibas;* Manobo—*agka-urepon*).

Ignorance of Indigenous Knowledge Systems

Poverty is also manifest in the loss of belief in and practice of the indigenous culture. Some possess limited knowledge while others stop learning and fail to cultivate indigenous knowledge systems (Manobo— *parabian-gantangan*). A distinction is also made between those who are poor in terms of indigenous knowledge and those who have such knowledge but fail to recognize and promote it.

Lack of Material Property

As in the outsiders' definition of poor, Lumads acknowledge that they may lack certain materials such as mats, animals, cooking pots, housing, food, and clothes (Ata Manobo—*ayuayu;* Mansaka—*gupisan/ allang;* Subanen—*dairun*). However, some Lumads insist that this may only be a temporary state due to factors such as migration to another localities and seasonality, e.g., before harvest (Subanen—*miskinan;* Manobo— *mangkaayo-ayo*). The Higaonon mention the *tingauhol* or seasonal hunger. During this period, they are forced to subsist on root crops and eat twice a day only.

THE ISSUE OF EXISTENCE OR NONEXISTENCE OF POVERTY

It is acknowledged that poverty as defined by Lumads does exist in every ethnic group. However, it is also reiterated that they have resources that the outsiders covet. Foremost, they respect richness in character (Manobo—*bungkatol ha bulawan).*

The Talaandig say there are no rich, so there are no poor. People raise crops sufficient for all, even enabling the community members to hold rituals. The resources are redistributed equitably by the local leaders *(timuay* or Datu) or there is a system of food sharing (Higaonon—*ilahan, pangalawat).*

According to those in Northwestern Mindanao, Lumads cannot be poor if the following are present.

- forests for all their food and health needs;
- peace;
- the indigenous characteristic of helping one another in times of need;
- knowledge about the forest in lieu of formal schooling; and
- indigenous systems and means to help them choose their own lifestyle.

Again, the conditions cited hinge on their access to resources such as land and forest, their cultural integrity, and the relationships fostered within and outside their communities.

NEW ISSUES AFFECTING INDIGENOUS PEOPLES

In the past, Lumad communities encountered difficulties from a limited number of sources, mainly from nature in the form of storms and the destruction of crops by pests. The peace situation has been so unstable that there is sometimes a need for evacuation of communities.

Now these communities have to contend with the entry of migrants from lowlands; the influence of religion, churches, and schools; and the use of technology by outsiders (e.g., in commercial farming). Other pressures include disruptive development projects such as dams, irrigation projects, integrated forest management and community-based forest management activities, contract reforestation projects, logging, mining, energy projects, and school reservations. In effect, there has been a lack of management and control of the territories occupied by indigenous peoples.

To aggravate the situation, Lumads contend with lack of knowledge in resource management because the indigenous knowledge systems and practices have been forgotten; and because of a lack of current technical knowledge to develop existing natural resources, lack of money, and inability to cope with the development standards set by outsiders.

Full Implementation of the Indigenous People Rights Act

Closely linked with the development of ancestral domains, the indigenous peoples seek the recognition of their rights to their domains, the right to develop the lands as they see fit, and to regain territories lost either through sale at a low price or acquisition through government agreements.

Education

Indigenous peoples see the importance of sending their children to school even if costs are beyond what they can afford. Thus, families do not hesitate to sell a piece of land or an heirloom to meet school expenses. They also forgo spending for agricultural production.

As in the Cordillera, those interviewed explained that a college degree for their children is their passport out of farming. Among the Higaonon, some have abandoned the traditional practice of arranged and child marriage in support of their goal for their children to complete high school or college education. Scholarships are welcome, although there is a clamor for more rigid screening to ensure that those selected really belong to indigenous communities.

The indigenous peoples, however, are also wary of negative influences of education in terms of the changes in values of their children. The loss of indigenous knowledge in return for lowland aspirations is a dilemma for these indigenous peoples.

Consistent with indigenous peoples' perception of the life they want to live are such aspirations as the full implementation of the IPRA, education, livelihood programs and financial assistance, self-governance, Similar to the NEDA survey findings on Minimum Basic Needs, the present study found that Mindanaoan indigenous communities aspire to have more income to provide for basic household needs like food, clothing, education, and medical services.

Like their Cordilleran counterparts, the most common livelihood activities are pig raising, fruit tree cultivation, vegetable production, handicrafts, and dressmaking.

Self-governance

Indigenous communities seek to manage their domains based on their own systems and methods. There are moves to assess the effect of outside influences on their culture and to reject what is incompatible and harmful to the local traditions and values. Moreover, those influences considered enriching could be modified and integrated into the culture and society.

There is a strong reaction to the entry of development projects without the benefit of sufficient community-wide consultation. Furthermore, if and when participation does occur, agreements are sometimes not respected or followed through. Also required is the implementation of free and prior informed consent in development matters.

Respect for the Indigenous Peoples and their Culture

In their everyday interaction with outsiders, the Lumads have to contend with the low regard of the former for the latter. There is an absence or a lack sensitivity and respect for indigenous knowledge systems and practices.

In this regard, the indigenous peoples want cultural revival and the promotion of their local systems to raise an awareness among outsiders of the richness of their traditions and to imprint indigenous knowledge, systems, and practices into the consciousness of succeeding generations. And revival of the concept of leadership such as the Datu and the *timuay.* The responsibilities and accountabilities, clear structures and functions, and the issue of transparency should be taken into consideration by both new and traditional organizations.

Improved Infrastructure

Indigenous communities see better access roads and bridges as motivators for increased productivity. However, they are also wary of the possible negative impacts in terms of the entry of individuals or groups with interests in their lands and resources.

In addition to roads, these peoples want to set up "tribal halls" (or centers where they can meet regularly), school buildings, water systems, and health centers. For infrastructure projects, they reiterate that their needs and not those articulated by external agencies (even those with genuine concern) should be given priority.

Participation in the Development Process

Increasingly, people from the communities want more meaningful participation in the development process. NGOs and government organizations working directly at the grassroots level have realized that the participation of community stakeholders in the conceptualization, planning, implementation, and monitoring/ evaluation of projects is key to the success of the projects.

Organization for Protection

The protection of the forests, lands, and people from outsiders is viewed as critical by the indigenous peoples of Mindanao. Given the pressures (e.g., political, financial, and developmental) impinging on their rights, there is a need to strengthen their traditional organizations and other groupings within the communities. The process would include the rethinking Several recommendations were put forward by the indigenous peoples of Mindanao with reference to the type of development efforts in which they will engage, whether processes, policy decisions, or projects. Some issues that need to be addressed by the NCIP deal with specific cases. These are the Central Mindanao University land claims and dismantling of the indigenous peoples' *tulugan* (the structure where the communities meet), problems with community-based forestry management (CBFM), and the need for a dialogue with the Office of Southern Cultural Communities (OSCC) regarding indigenous peoples (e.g., the Teduray) within the ARMM. The most frequently recurring issues raised during consultations with indigenous peoples refer to their lands. This may take on the form of landgrabbing by powerful politicians and business persons, entry of CBFM, tourism, plantations, projects such as dams, and individual claims by nonindigenous persons.

CASE STUDY 1: THE HIGAONON COMMUNITY IN BUKIDNON

General Demographic and Socioeconomic Situation.

Access and population

- Translation of the Policy on Indigenous Peoples into local languages (e.g., Visayan) to be used as a reference in discussing the parameters of engagement with ADB.

- Support in the processing of CADTs with NGO assistance; the latter need to be acceptable to the communities assisted.
- Assessment and nonfunding of projects which, based on experience, are not supportive of indigenous peoples' rights; some of these are CBFM, mining, integrated forest management, power plants and other energy projects, dams, major infrastructure, and fisheries/aquatic resource management projects.
- Participation of the indigenous peoples in the review of proposals for projects affecting their communities. Funding should not be provided without their FPIC.
- Creation by ADB of an ad hoc independent investigation team (with an indigenous peoples' representative) to review, monitor, and evaluate ongoing projects within their domains.
- If possible, channeling of projects straight to the indigenous peoples' communities.
- Continuous consultations with indigenous peoples.

Considering the current conflicting policies impacting on indigenous peoples, ADB should apply

Accessible mainly by the *habal habal* (motorized bike that accommodates at least four individuals including the driver) from the highway, the Higaonon share their 6 villages (5 of which are within the 9,000 hectare CADCs) with 10% Visayan migrants. There are 88 households (1995) of indigenous peoples and *kalibog* (persons of mixed ethnicity) in the community.

Land cultivation and access

The Higaonon traditionally practice shifting cultivation, hunting, and gathering of forest products. They raise upland rice, root crops, sugarcane, and (in the 1970s) coffee and different types of vegetables. Recently, they have started to practice settled agriculture, using high-yielding varieties of rice and corn.

Since the prohibition of swidden farming (PD 705), the residents' movement over their traditional territory has lessened. Although the awarding of the CADC has given them some sense of security of tenure over their land, they profess confusion about the requirements of the municipal government regarding the paying of community taxes and municipal assessments conducted annually. They expect that ownership over their territory will be fully recognized and respected through the issuance of the CADT.

Since the Higaonon have been slowly converting from swidden farming to lowland agriculture, construction of rice paddies is perceived to be a symbol of promise of food security and better life because it will enable them to plant and harvest thrice yearly. It is also construed as a strategy for forest conservation since they would no longer clear some portions of the forests. The shift in agricultural production has been strongly influenced by the government's forest policies, the entry of migrants, the Municipal Agricultural Office, and the Catholic Church-based NGO Father Vincent Cullen Tulugan Learning Development Center (FVCTLDC).

The last mentioned group has provided various forms of assistance to encourage the shift to lowland agriculture. However, support for inorganic inputs was later stopped because of the perceived detrimental effects on health and environment.

Employment and incomes

Cash is generated from the sales of excess agricultural produce as well as raw and processed abaca hemp (woven cloth, *hinabol*) and rattan (backpack, *kamuyot*). For these products, rates are dictated by entrepreneurs who transport and sell products in the town centers. Some NGOs have been providing support to construct a building for weaving looms and a market outlet for products of Higaonon women. From their initial income, the women were able to set up a cooperative store at the center of their village.

Another livelihood opportunity is employment as government officials, soldiers, and Citizens Armed Forces Geographical Unit members. Some residents leave the community seasonally to seek work as farm laborers and domestic helpers. The average income from farm labor is P70–80 per day for adults and P50 for children for 8 hours of work. Cash is used for basic household needs such as rice, salt, sardines, kerosene, candles, matches, and soap.

Education facilities

Educational facilities consist of two public elementary schools, one in the town proper and another in one of the villages.

Secondary education has to be pursued outside the area. The FVCTLDC runs a literacy school in one of the villages. It offers a special curriculum including subjects on Higaonon culture. However, the institution still needs to be accredited by the Department of Education, Culture, and Sports; thus, students must undergo placement examinations to qualify in the regular school program.

Parents opt for their local language or Visayan for instructional purposes since these are easily understood. However, this does not diminish their openness to learn the national languages, Tagalog and English, because these are viewed necessary in communicating with outsiders.

Health

To respond to the health needs of the community, a health center in the town proper has one midwife offering mainly family planning services. The FVCTLDC also runs a health clinic with a registered nurse. There are plans of building a minihospital within the town by the ADB-funded Bukidnon Integrated Area Development Project.

Not all villages possess a water system; the government has installed intake boxes that are not yet operational. Consistent with their objectives, the FVCTLDC helped build a water system in one of the villages.

Through the inputs of the FVCTLDC among women, health and sanitation in two villages has improved, as indicated by zerc infant mortality in 2000 and the prevention of epidemics. The Center encourages the use of traditional and indigenous medicine and these are prescribed rather than "western" medicine. The Higaonon still avail of the services of the indigenous medical practitioner after consultation at the health centers.

Political institutions

The Higaonon continue to practice, recognize, and respect the traditional Datu leadership system. Although some of their members have been elected to various government posts, traditional leaders claim harmonious relationships with the latter.

Recognized in the area are two main Datus and others who comprise the Council of Elders. The Datus have specialized skills recognized and respected by the community. Collectively, they take charge of ceremonial functions, conflict resolution, and healing functions. In effect, they make the major decisions concerning the community.

Recently, tensions and undue competition emerged among Datus from different sectors because they believe that some leaders have greater access than others to government and NGO projects. Village leaders and members perceived to have more access to NGO projects are not regularly invited to government-initiated meetings, consultations, and planning activities, and their areas have been less prioritized for development projects and delivery of basic services.

Conflicts within the ancestral domain are resolved by Datus while issues affecting areas beyond the ancestral domain are jointly resolved by the Datu and the village captain.

Beyond the community, all the Datus in the area are linked to a traditionally recognized and informal organization of leaders from different Higaonon territories in Bukidnon, Misamis Oriental, and Agusan del Sur. They assemble from time to time in a ritual called *dumalungdong* to discuss their customary laws, belief systems, and pressing problems.

Concept of Poverty among the Higaonon Terminology and definitions

When asked for local terms for *poor*, the response would be Visayan terms *kabos* and/or *makaluluy-on*, although the latter is more linked to being in a pitiful condition. The term can be equated with the absence or lack of money while its opposite, *sapian* (Visayan for rich) connotes the ability to purchase material things such as houses, farm animals, and vehicles. These terms were not within their consciousness until the entry of outsiders.

Thus, they perceive themselves to be poor only because others say they are poor. One informant refused to be referred to as pitiful because she still has food on the table and does not have to resort to begging.

The nearest terms akin to poverty are *agkapulog, tingauhol, and kalugan.* Residents experience the first only because they do not have money to buy certain things they need, but this condition does not necessarily indicate that food is scarce. The second term refers to seasonal hunger. Although food is limited during this period, the situation is temporary and can be mitigated. The third term means a person in a state of difficulty: while walking through a steep and narrow terrain, when food is scarce, when a conflict is unresolved, or when violence is experienced.

Poverty promoting factors

The Higaonon perceive themselves to be in difficult circumstances caused mainly by outside factors. According to ethnographic literature, they were traditionally coastal dwellers enjoying the resources both of the sea and the forests. However, they were subsequently pushed into the interior areas and had to depend solely on the forest for subsistence and household, health, and ceremonial needs.

The interior areas where they settled were then encroached upon by migrants in the 1970s. A number of Higaonon lost tracts of land to outsiders for cash or for cheap goods.

The situation was aggravated by government policies on land tenure and resource use, greatly limiting access to forest zones. Then logging activities by the Nasipit and Agusan logging companies from 1980 to the early 1990s affected the economic, social, and political aspects of the community.

With the depletion of forests, the more immediate impact was the decrease of meat supply (deer, wild pig, monkey, chicken, frogs, birds, and fish). In addition, sources of honey became difficult to find. The *manuk gasanun* (wild chicken) for traditional healing practices became scarce.

Logging also greatly altered the cycle of traditional farming systems. Due to more intensive use of the land, farm productivity lessened. The problem was aggravated by a build–up of pests and disease. As a result, the harvests have not been sufficient to sustain the Higaonon communities between croppings.

As the supply of traditional food dwindled over the years, seasonal hunger became more prolonged. Consumption patterns also changed: residents learned to eat canned foods and instant noodles. These are nutritionally inferior to their traditional diet of vegetation and animals from rivers and forests.

Needs and Aspirations

The Higaonon needs and aspirations include infrastructure, production, social services, and the conservation and promotion of indigenous systems and governance.

- Infrastructure and facilities: better roads and transportation systems, electricity and a vehicle for the community.
- Production and related facilities: construction of more rice paddies; production of more root crops, corn, abaca, sugarcane, and coffee; setting up a rice and corn mill; construction of productive fishponds; identification of markets for farm produce.
- Social services and support systems: support for education, fair and just gender distribution of labor, and strengthening of the cooperative system.
- Indigenous systems: preservation of indigenous beliefs and value systems (especially values of helping, caring, understanding, and respecting one another); continued performance of traditional rites and rituals (guidebook on rites and rituals); and exclusion of new religions into the area.
- Governance: continued strengthening of traditional political organization and structures; recognition of the youth as community leaders, both in the traditional and state mandated systems; and learning new knowledge and skills to help strengthen community organization.

Currently, the Higaonon community is optimistic that with the financial and technical help of the FVCTLDC and its network of NGOs and other support groups, the identified needs will eventually be met.

CASE STUDY 2: THE ATA-MANOBO COMMUNITY IN DAVAO

General Demographic and Socioeconomic Description

Characteristics and location

This group is one of three Ata Manobo "subgroups." Its territories, part of which is in a CADC area, stretch from Davao City to Bukidnon. This may be due to the fact that some members are still nomads who move around specific domains. The town proper is accessible by public vehicles, mainly motorized bikes, *habal-habal.* The latter serve as the only means of transportation to reach the interior communities. However, within some villages, horses are used to transport people and goods.

The area is known to be rich in minerals such as chromite, gold, silver, and coal. Water still has to be fetched from its source (e.g., springs); there are plans for spring development leading to a water supply system. According to a respondent, electricity is enjoyed only by the former mayor (a Datu) and his family from a generator at the municipal hall.

Production and other subsistence/ livelihood practices

The Ata Manobo are swidden farmers in areas called *lugar na hagpo.* The products of this group are root crops such as sweet potato, cassava, and taro. Also grown are corn, banana, coffee, and abaca. Currently, an added product is upland rice. Problems faced by the farmers include rats that feed on the harvest and lack of capital or financing for farm inputs.

There is a traditional cooperative labor exchange in which families provide assistance to one another.

Women weave cloth (*hinabol*) and mats as well as baskets and containers from bamboo and rattan. Some men are blacksmiths and make their own farm implements. Some still hunt using a bow and arrow. Others raise animals such as chickens, pigs, and horses.

The income of the Ata Manobo is hard to estimate because of irregular marketing schedules and volume of products. The most common item used to generate cash is wild abaca transported to the market in bundled strings. Corn is likewise sold through entrepreneurs who serve as credit sources.

Employment is in the form of farm labor in the village or migration of children to other places to work as house helpers. For the former, about P3,000 is generated per harvest while house helpers are paid P400 per month.

Education

During 1977–1984, the Presidential Assistance for National Minorities provided schools for elementary education and scholarships. Currently, access to education has become a problem.

Usually the family borrows to finance the education of the children using the harvest as collateral.

Parents want their children to be taught according to the language of the lowlanders although a curriculum for Lumads has been developed. Teachers are composed of both Lumads and Visayans.

The Davao Medical School Foundation, an NGO, has provided scholarships for midwifery. A day care center is also present but the structure has not been used due to the limited number of users.

Health situation

The Municipal Health Center has a nurse, midwives, and village health workers. The staff seldom visit the interior areas to deliver health services. Women healers (*oyami*) provide childbirth assistance and treatment of diseases, while family planning is conducted through the use of indigenous herbal contraceptives.

Other indigenous practices

Very few Ata Manobo reject their traditional religious beliefs. Some, however, have become Catholics, Seventh Day Adventists, and Baptists. But it is said that they eventually return to their traditional practices.

Although indigenous peoples are subsistence based, they also access credit from individuals. Loans must be paid or shame will befall the Datu. The people do not want their Datu to be dishonored, so they pay their debts.

The Ata Manobo are said to share property. Even a radio may be passed from house to house in the spirit of sharing.

Assistance provided to the community

The community has a CADC and the contract for the formulation of the Ancestral Domain Management Plan was awarded to Multi-Equipment and Technical Services, Inc., a private corporation affiliated with the Marsman Company. The budget was generated from the Poverty Alleviation Fund (PAF III). This source also provided for the labor of residents in four related activities: brushing, staking, hole digging, and hauling of seedlings. Conflict erupted in 1999 because residents who provided labor were not paid, according to one of the Datus, and there were apparently irregularities in contracts.

In addition, a former government program, the *Lingap para sa Mahirap,* targeted 19 families as the "poorest of the poor." A cooperative, a sari-sari store, and a loan from the Davao Tree Farmers' Cooperative were planned but had to be shelved because of the limited skills level of residents in financial management.

Other government entities that have provided assistance to the group are the Department of Agriculture (which provided 12,000 coconut seedlings since 1995) and NCIP. The latter conducted leadership training and assisted in negotiations between the Datus and a logging company.

The Davao Medical School Foundation, an NGO engaged in health, education, and livelihood assistance provided community workers to work in the villages. They implemented such projects as education assistance funded by Save the Children Japan, and goat raising. The 6 goats distributed to different households have bred and now number 26.

The Nestlé Company provided farmers with training on coffee planting. The villagers, although interested, failed to use the free seedlings because of problems regarding transport.

Conflict resolution and the Datu system of governance

The Datu is the center of governance and conflict resolution in the Ata Manobo communities. Common issues are murder, adultery, theft, and unpaid debts.

The Datu should have goodness and strength of character, be helpful to the community in development plans and services, generate respect and trust from the group, and look after the peace in the community. The Datu's power is inherited but there is also a selection process to identify an individual worthy of the position among offspring and close relatives. The Ba'e or wife of the Datu is also respected and has important leadership functions.

Peace and order situation

In 1972 the communities complained of abuse by the military; troops were subsequently withdrawn. A volunteer police force has since maintained peace and order.

As of April 2000, it was said that about 40% of the approximately 100,000 inhabitants of the area are affiliated with the New Peoples Army, an antigovernment military force. These people reside within the CADC.

CONCEPT OF POVERTY AMONG THE ATA MANOBO

Terminology and definitions

The term akin to *poor* among the Ata Manobo is *kaayo-ayo,* an individual described as possessing only a set of clothes and who has to live with others for his/ her subsistence.

In a worse condition would be the *kaubong-ubong,* one who is practically without any living area and who does not attend to his/ her physical appearance (e.g., does not take a bath). Individuals are said to be in this state because they reside in the interior areas and do not participate in community activities.

The comfortable ones, on the other hand, are the *maupiyak.* They have sufficient food and domesticated animals. This perception of a level of comfort as stated by the Ata Manobo contrasts with outsiders' views of the wealthy *(adunahan):* possessing much money, concentrating on making a business prosper, and pursuing only his/her personal interest.

Poverty promoting factors

As recently as 1983, the forest in the area was still intact. The Datus interviewed said they were living in harmony with the forests, which produced all they needed. Then, a 25-year logging concession was granted to the Alcantara company. The company has been in the area for 15 years. There was a steady depletion of forest areas and loss of biodiversity. Wild animals vanished and farmlands were taken over by the logging operations.

Naupaw ang lasang tungod sa logging sa Alcantara. ("The forests turned bald due to the logging of the Alcantaras.")

Now, the people cannot clear the cogon grass areas. There is insufficient organic matter due to lack of forest litter. The cultivation of upland rice has become difficult because of the limited amount of planting materials and the soil has become too compact, making even sweet potato cultivation difficult.

At first logging companies cut the trees, then they reforested the area with fast-growing trees and prohibited the Lumads from farming within these reforested areas.

These factors have resulted in food being the primary problem in the area, even if the Poverty Alleviation Fund gives food for work.

The logging company tried to help by implementing projects: a drier, a water system, and provision of seedlings. New methods and technologies, however, required the introduction of new methods that needed money to implement.

Changes were also noted by the Ata Manobo in their cultural practices. Previously, they could work together even without money through the *lusong* or cooperative work. Now there is the concept of hired labor. In addition, they now fail to initiate rituals before they open the land and they now use metal tools in clearing farmland.

The conditions are aggravated by the presence of lowland Christians who are said to oppress them by buying products cheaply and selling them at exorbitant prices. In addition, they have also brought in vices such as drunkenness and gambling.

9

Review of Development Programs and Projects

The programs and projects reaching the indigenous peoples of the Philippines are made possible through the support of international agencies, the government, NGOs, and IPOs. Presented here are samples of these endeavors and suggestions for ADB to consider.

INTERNATIONAL AGENCIES WORKING WITH INDIGENOUS PEOPLES

The various international agencies engaged in work with indigenous peoples of the Philippines are guided by their own policy frameworks. The best measure of the effectiveness and relevance of their strategies is a review of their articulation at various levels: the agency, the host government, the partner NGOs and IPOs, project staff, and the indigenous peoples themselves.

International Labour Organization

The International Labour Organization (ILO) serves as the only international agency that has adopted specific standards covering indigenous peoples. To address the issue, ILO has taken a twofold approach of promoting its Convention 169 at the government level and providing technical assistance on indigenous peoples issues.

ILO's current development program for indigenous peoples is INDISCO, which uses a community-driven participatory approach rather than a top-down approach. The latter was found ineffective since it makes indigenous peoples dependent and is considered a disservice to their communities, which were traditionally self-reliant. INDISCO's current projects are classified into four components: preservation and promotion of the indigenous culture, promotion of decent employment and income opportunities, promotion of gender equality, and environment and natural resources protection and rehabilitation. There are 4 indigenous communities involved in these projects in Mindanao, 2 in Occidental Mindoro, 2 in the Cordillera, and 1 in Zambales.

United Nations Development Programme

The Global Environment Facility (GEF), established in 1990, is comanaged by the World Bank, UNDP, and the United Nations Environment Programme. It is designed to assist developing countries to deal with four main environmental problems: global warming, pollution of international waters, destruction of biological diversity, and depletion of the stratospheric ozone layer.

The Small Grants Programme (SGP), which is managed by UNDP under the GEF, supports community-based activities that address the above mentioned concerns.

In the Philippines, the objectives of the GEF-SGP sponsored projects are the following.

- Promotion of biodiversity conservation in selected/priority biogeographic regions.
- Protection of international waters from environmental impact of activities from within the territorial boundaries of the Philippines.
- Reduction of greenhouse gas emissions through the promotion of renewable energy in rural areas.

Since 1994, the GEF-SGP has approved 21 projects that involve NGOS and indigenous peoples. Of these, 16 focus on biodiversity conservation, 4 on climate change and 1 on depletion of the ozone layer. In terms of location, 12 are in Luzon, 3 in the Visayas, and 6 in Mindanao. In biodiversity conservation, they include a variety of studies of traditional systems and conservation of wild flora.

Philippine-Australia Community Assistance Program

The Philippine-Australia Community Assistance Program (PACAP) is one component of Australia's Development

Cooperation Program. It supports community-initiated, economically sustainable, ecologically sound, and gender responsive development efforts. To achieve these goals, PACAP provides direct funding assistance and multilevel capacity building in cooperation with NGOs and people's organizations with the goal of raising the living standards of poor and marginalized Filipinos.

PACAP started in the Philippines in 1986 and to date, it has supported 932 projects. Of these, 95 projects involve indigenous peoples and 7 were ongoing in 2001. These are on agricultural development including crop production, postharvest facilities and marketing, and management of the environment and natural resources on a sustainable basis (including soils, water, fisheries, habitats, and forests). Six of the projects are in Mindanao and one in Baguio City.

The World Bank

The World Bank's Operational Directive 4.20 guides Bank projects dealing with indigenous peoples. It ensures borrower government adherence to these standards. Currently, consultation efforts are ongoing to finalize the revision of the policy.

In the Philippines, the Bank has a total of 24 projects. Seven refer directly to indigenous peoples and a further 14 are implemented in areas where indigenous peoples reside. Another project, the Land Administration and Management Project, will benefit NCIP because it deals will the systems and processes necessary for ancestral domains delineation and titling. Projects implemented in areas with indigenous peoples deal with women's health, education, rural finance, agrarian reform, water, development of local government units, social expenditure, and a transmission grid.

Those with direct reference to indigenous peoples are the following: Early Childhood Development, Community-based Resource Management, SZOPAD Social Fund, Conservation of Priority Protected Areas Project, Mindanao Rural Roads Development Project, National Roads Improvement and Management Project, and Subic Bay Freeport Development II.

In 2000, the World Bank provided a grant to NEDA to undertake a project entitled Capacity Building for Social and Environmental

Assessments. The project aims to strengthen the capacity of government project staff and implementing and oversight agencies of government, universities and colleges, and NGOs and people's organizations to undertake work in Bank-financed projects. These are in the areas of social and environmental assessments, and the preparation of resettlement and indigenous peoples' action plans. The project is ongoing.

The operational guidelines set by the World Bank serve as the reference for the Philippine Government in dealing with indigenous peoples' concerns. However, the articulation of the guidelines at the ground level, according to project-based consultants, depends on the approach of the contracted staff. An example is the concept of participation in the formulation of the Ancestral Domain Management Plans. While one consultant conducted training among the indigenous peoples for the latter to produce the said Plan, another consultant hired an anthropologist to conduct a study and write it. It is critical for organizations to review how policies are carried out at the community level in the reformulation of strategies and, especially, policies.

MAIN GOVERNMENT PROGRAMS AND PROJECTS

The recent draft of the 2001–2004 Medium Term Philippine Development Plan considers poverty reduction as its overarching objective. According to Secretary Deles, for the first time in the history of the Philippine development plans, there is a chapter on Mindanao where most of the indigenous peoples reside.

The main strategies for economic development and social reconstruction hold the promise of changes in the lives of indigenous peoples if they are pursued with a strong political will by the Government. The strategies are as follows.

- Relief and rehabilitation in conflict-affected areas. This will entail relocation and resettlement of families.
- Provision of basic services, livelihood opportunities, and other needs that would ensure minimum desirable levels of welfare. Even psychosocial needs will be considered.
- Implementation of culturally-sensitive programs. Health care approaches will entail intersectoral collaboration to make them sensitive and conform to cultural norms and practices. Traditional health workers will complement the

regular government workforce. For education, indigenous concepts will be integrated into the mainstream educational system. Basic curricula for Muslims and indigenous peoples will be revised to make them more relevant.

- Institutional reforms to address political inequity. The participation of Muslims and indigenous peoples in decision making will be broadened.
- Implementation of the IPRA. The focus will be on the return of ancestral lands to indigenous peoples for their utilization.
- Full implementation of the 1996 Government-MNLF Final Peace Agreement.
- Peace talks with the MILF.
- Peace education and advocacy.

It is evident that an arena for engagement between the Government and the indigenous peoples has been set as specified in the new Philippine medium-term plan. Line agencies have been given their mandate and are in various stages of articulating this in their projects and programs.

The following government line agencies already have clear programs responding to the needs of indigenous peoples in communities where services are delivered.

Department of Health

Recently, the Department of Health (DOH) issued departmental policy No. 142, mandating provision of resources for indigenous peoples' concerns. Moreover, DOH has created a National Technical Working Group for an Indigenous Peoples Health Development Program. Another noteworthy initiative is the creation of the Indigenous Peoples Health Development Project-Partnerships, which include the provision of grants for capacity building and community projects for local NGOs and peoples' organizations in partnership with indigenous peoples' communities. The projects include water system development, nutrition, community development training, environmental health sanitation, and prevention and control of local endemic diseases. The partnerships program is currently working with the Dumagat in General Nakar, Kongking in Ormoc City, Bihug in Camarines Norte, and Agta in

Central Luzon. DOH focuses on training volunteer health workers from the communities with emphasis on their roles as community mobilizers, health service providers, and health learning facilitators; advocacy and promotion of effective health strategies that directly benefit indigenous peoples; and strategic health planning workshops for indigenous peoples. For 2000 to 2001, the proposed activities included the following.

a. Needs assessment to identify priority areas for assistance.
b. Construction of 8 village health structures in the 7 provinces/cities of Region XI.
c. Capacity building of service providers in eight indigenous peoples' areas in Mindanao.
d. Development/improvement of 12 water sources to level I and level II in indigenous peoples' communities in Mindanao.

Department of Social Welfare and Development

The Department of Social Welfare and Development is currently implementing the Comprehensive and Integrated Delivery of Social Services (CIDSS), one of the 10 flagship programs of the Social Reform and Poverty Alleviation Act (RA 8425). Its goal is the empowerment of disadvantaged families, communities, and sectors so that they can access basic services and manage their resources. One of its funding sources is the Australia-funded Vulnerable Group Facility. Currently, CIDSS is being implemented in 1,084 municipalities, 3,793 villages, and 100 poor communities.

For 2002, a CIDSS program to Liberate the Indigenous Peoples from Indignity is being carried out. It seeks to extend culturally appropriate projects to indigenous peoples' communities in 100 CADC villages to ensure their survival; build their capabilities in managing community-based services; preserve their ancestral domain; and cultivate their cultural heritage through the convergence of government, nongovernment, and other sectoral resources, expertise, and technology. It will be jointly implemented by several government agencies.

The objectives of the program are to establish an ethnographic profile; strengthen indigenous values, knowledge systems and practices through rediscovery and reorientation sessions and needs assessment tools; provide basic services to the communities; and modify/ develop and institutionalize culturally appropriate projects and programs for families and individuals/ sectors in need of special protection. A few activities address sustainable upland socioeconomic development through improved agricultural technology, access to basic services and markets, and strengthening existing community organizations.

Department of Education, Culture, and Sports

At the national level, the Department of Education, Culture, and Sports is embarking on two major activities to become more responsive to indigenous peoples' needs, specifically in preserving their cultural heritage: to indigenize elementary and secondary curricula and to develop an equitable nonformal educational system.

A project entitled Basic Education Assistance for Mindanao covers Regions XI, XII, and ARMM. Its objectives are to expand educational opportunities for the poor and disadvantaged population groups, promote curriculum flexibility and localization, and strengthen partnerships in education. These objectives are to be met by establishing 80 community multipurpose and learning centers in remote communities; providing scholarships for selected poor Muslim and indigenous peoples high school graduates to complete teacher education courses; and setting up cultural enhancement facilities in 18 strategically located schools, with significant enrolment of indigenous children.

Department of Environment and Natural Resources

The DENR regular activities that involve indigenous peoples are the Management of Ancestral Domain Claims and the Community-based Forestry Management Program.

With funding support from international agencies, DENR is currently implementing projects to address IP concerns, including biodiversity management and conservation, and forest and watershed management with emphasis on community participation and self-The Republic of the Philippines-Asian Development Bank Poverty Partnership Agreement (10 October 2001) specifies the need to "provide adequate protection for the poor, vulnerable and disadvantaged sectors." Further, it indicates the necessity of setting up community and home-based social service interventions using local government units, NGOs, peoples' organizations, and cooperatives as delivery mechanisms.

Although these NGOs and IPOs may not always explicitly refer to the term *poverty reduction* as their goal or objective, the activities and expected outputs point to the concepts of participation and empowerment, sustainable development, and improved quality of life among the marginalized sectors. These concepts are explicit in the projects dealing with delineation and recovery of ancestral lands, health, education, livelihood, and technology transfer. All these address poverty-related issues and, if effectively implemented, should result in a better quality of life among indigenous peoples.

These NGOs and IPOs may operate at a national level (with federations), but would have a specific constituency (e.g., area-specific IPOs), priority locations (e.g., mostly Mindanao), and field of expertise (e.g., ancestral domain delineation) for project implementation purposes. There is also a wide variety of affiliations (e.g., the church, the "left", academe) and, thus, sociopolitical orientation guiding the method and direction of projects and programs.

A preliminary listing of relevant NGOs and IPOs was generated from the 7 ethnographic area consultations by OPAIPA in 2001. NCIP is expanding its inventory of such organizations for purposes of networking, depending on their geographic location, priority projects, and skills/specializations.

Priority Thrust: Ancestral Land and Domain

Currently, there are 181 approved CADCs and 9 CADTs proposed by NCIP. The Commission has no funds for land delineation and related activities, necessary equipment, or staff capability; it has a limited quantity of documents turned over by DENR. Thus, NCIP relies on private surveyors and NGOs with expertise in delineation and has made memoranda of agreement that give these groups (e.g., the Philippine Association for Intercultural Development [PAFID] and Anthrowatch) the authority to deal with mapping and delineation.

PAFID has the most expertise and number of communities assisted. Since 1974, the organization has participated in the campaign for tenurial and self-management aspects of ancestral domains in more than 600 indigenous communities and has completed work on up to 85 CADCs and CADTs. It has been conducting training since 1994 within and outside the Philippines. In 1995, DENR and the Bureau of Coasts and Geodetic Survey formally recognized PAFID's outputs. Based on its track record, PAFID is called upon to assist in solving urgent issues that affect indigenous peoples. These are referred to as "quick response" cases, the latest of which was the killing of some Manobo while delineating their lands and the burning of a village in Mindanao.

Other groups undertaking work in this field are the Green Forum, the Mangyan Missions, and Anthrowatch.

Anthrowatch, organized in 1994, is building equipment for mapping and delineation, and works with PAFID in Romblon, Panay, and parts of Mindanao. It also conducts community organization and capacity building, cultural sensitivity training, research on the health situation among indigenous peoples (with DOH), coordination of multistakeholders' meetings/ discussions, and publication of materials relevant to indigenous peoples' concerns. To this end, it facilitated a study tour by indigenous peoples for exposure to similar communities with a high level of organization.

PANLIPI and the Legal Resource Center handle the legal aspects of land issues. Both NGOs are active in advocacy and consciousness raising/education with reference to the IPRA. Environmental Science for Social Change Inc. also does surveys and makes maps. It is active in mining and plantation areas. The Upland NGO Assistance Center provides technical and academic support in dealing with indigenous peoples.

Formulation of ancestral domain management plans and ancestral domain sustainable development protection plans involve NGOs such as *Sentro ng Ganap na Pamayanan* and the *Ugnayang Pang Agham Tao* (UGAT or Anthropological Association of the Philippines).

The *Pambansang Lupon ng mga Nakatatanda sa Tribu,* an IPO organized as a result of a PANLIPI program of action for their organization's Council of Elders in 1994, has a total of 40 member organizations in Central Luzon, Northern Luzon, Mindoro, Palawan, and Mindanao. It undertakes paralegal training on issues related to ancestral domains and undertakes strengthening of IPOs. Another IPO, Tribal Education for Ecological Systems Inc. in Compostela Valley, Mindanao, assists in self-delineation. Other IPOs like *Katutubong Samahan ng Pilipinas, Panagtagbo,* and Lumad Development Center conduct most of the organizing and groundwork before mapping activities begin.

Issues in Working with Other Agencies in Land Issues

The difficulties encountered deal with orientation of and competition among NGOs. Those on the "left" do not value the process of land titling within the framework of the IPRA. Criticizing the law as "deceptive and divisive," organizations identified with the left (e.g., Cordillera Peoples' Alliance and *Kalipunan ng Mamamayang Katutubo ng Pilipinas)* prioritize community organizing, advocacy, education, and mobilization campaigns on issues and concerns of indigenous peoples. Some NGOs and IPOs have "claimed" certain indigenous communities as "their areas" and usually monitor the entry and activities of "outsiders" and can prevent the movement of individuals conducting mapping activities. Several other NGOs have taken advantage of the "space" provided by the IPRA to advance indigenous peoples' rights to their ancestral land/domains.

Interaction with the Government is facilitated through professional ties established with individuals within NCIP, DENR, and the National Mapping and Resource Information Authority (NAMRIA). There remains the need to work formally with the Housing and Land Use Regulatory Board and the Department of Agrarian Reform. Their participation will facilitate delineation at the ground level.

It is said informally that continued mutual openness between civil society and NCIP will help advance IPOs' goals, and that checks and balances should be set up by NCIP. Sincerity coupled with technical accuracy is critical because it has been observed that "more indigenous lands were lost through mapping than through bullets."

Lessons Learned

Based on the experience of the NGOs and IPOs in ancestral land and domain issues, the following aspects merit attention.

Professionalize. Once a decision has been reached by an NGO to undertake mapping and delineation, a unit should be set up. The staff should be trained and they should focus their work on the task at hand. Expertise is built up in this manner. A combination of experience and training among NGOs, IPOs, government organizations, and communities further hones the skills of staff.

Network. There is a wealth of knowledge in the various local and international agencies. Thus, formal linkages should be developed with these groups.

PAFID's first attempt at on-ground mapping of 3,980 hectares in Mindoro took 3 years (1989–1992). A series of linkages accelerated their pace and expanded the extent of their coverage. In 1993, the Ford Foundation introduced the group to the expertise of Dr. Urayvan in Thailand.

The World Wildlife Biodiversity Program provided a geographic positioning system and the Foundation on the Philippine Environment facilitated training on its use. Three trainers sent by the Washington-based manufacturer codeveloped a curriculum with PAFID.

Clark University in the US involved PAFID (in a joint project among 11 partners globally) in developing its data management software for NGOs. In 1995, during an international training event, the concept of ancestral waters was shared by the Mesquite Indians of Panama and successfully implemented by the Tagbanua of Palawan.

Moreover, visits and communication by individuals and organizations serve as catalysts in the continuing learning process. The US Peace Corps conducted extensive documentation of the Mindoro experience. An Australian volunteer brought in her knowledge on geodetics. Another volunteer, a US trainee with expertise in remote sensing, wrote a manual with PAFID.

Invest. Equipment should be bought and upgraded as necessary. Its use should be optimized and the system analyzed such that repairs can be performed by staff. There should be continuous links with experts to upgrade skills.

Train. PAFID has trained 127 indigenous people on the whole process of delineation from the initial survey to validation. One

Dulangan Manobo from Sultan Kudarat has achieved sufficient expertise to conduct training in other countries.

Trust and respect. Participation and partnership are critical in delineation. The indigenous peoples take care of securing the team and the instruments (handling and safekeeping of hardware). Their knowledge of the terrain and the people promotes ease and speed in the flow of operations.

There is sharing of tasks in community preparation and in mapping activities.

Thus, it is important that NGOs listen to the community and explain clearly the concepts and tasks for completion of delineation activities. When translated to cost, the residents as counterparts reduce the cost of the project because, in addition to all the other inputs, they provide labor and food.

Mutual trust and respect form the core of the engagement. As PAFID puts it, the activity is "not just mapping but it is also making a statement regarding indigenous peoples' lives."

Prospects for the Future with Reference to Ancestral Land/ Domain Issues

A workshop could be conducted to consolidate existing efforts and knowledge among various organizations and agencies. NGOs have equipment that their staff are unable to use. Some hardware is lying idle.

There is a need to tap existing resources such as Environmental Science for Social Change Inc., Geoplan (a Department of Science and Technology-funded NGO), the Mangyan Missions, and the geographic information systems laboratory of the University of San Carlos, Cebu City.

An outcome of such a workshop could be the setting up of a resource management institute or center where

- all endeavors could be coordinated, integrated, and housed;
- basic training for NGOs, IPOs, and other interested parties could be conducted; and
- networking and skills and hardware development could continue.

Response to Other Issues Confronting Indigenous Peoples

As noted earlier, there are many NGOs and IPOs dealing with issues affecting indigenous peoples. The main concern of most is ancestral land and delineation, but some embark on other endeavors as required by their constituency. Some examples are cited below.

In the area of education, NGOs such as the Asian Council for Peoples Culture advocate the creation of indigenous schools (culturally responsive primary schools for indigenous children). The Kalahan Foundation in Santa Fe, Imugan, has been conducting education sessions with the Ikalahan and Mangyan.

Sentro ng Ganap na Pamayanan is currently conducting a policy review and analysis, and documentation of indigenous knowledge for proposed integration into the government's education curricula.

General assistance is provided by the Episcopal Commission on Indigenous Peoples in 16 dioceses and to 40 active IPOs in the Philippines. The group also serves as a national secretariat and coordinates policy advocacy in the three branches of the national Government —executive, legislative, and judicial. It has been active in relief and rehabilitation work with the Tabang in Mindanao.

The National Confederation of Indigenous Peoples in the Philippines is composed of federations of IPOs and was organized in 1996 as a result of an ILO-Danish International Development Agency project managed by *Sentro* in response to the need for livelihood opportunities among indigenous peoples. Employing community organization, the staff, composed of indigenous peoples themselves, provided technical and organizing assistance to indigenous peoples' groups.

In 1998, it was formalized as a confederation. Among other objectives, it aims to be effective in the assertion, protection, promotion, and enjoyment of the indigenous peoples' rights already enshrined in international instruments, national laws, and policies, including those formulated by local organizations.

Linking with the established political base is one area where the Tribal Communities Association of the Philippines is active nationwide except in Muslim areas. It has strengths in advocacy, in identifying problems on the ground, and linking with local authorities and agencies to address problems of indigenous peoples.

Moreover, it provides legal assistance to indigenous peoples charged with criminal activities such as illegal logging, and in resolving conflicts arising from claims to ancestral domains. The Association took part in recent elections as a Party List candidate to bring forth the issue of the marginalization of the indigenous peoples in the political process.

The experiences of NGOs and IPOs in the fields of community-based education, health, livelihood, and sociopolitical activities/programs are indeed valuable. The challenge is how to tap their expertise and coordinate their initiatives so that they can contribute to development of ancestral domain sustainable development and protection plans.

10

Asian Development Bank Experience

In May 2001, some 150 Tinguian elders of Lamao village, Bucloc, Abra, gathered in a community meeting arranged by the community mobilization officers of the Cordillera Highland Agricultural Resource Management Program (CHARM). The elders were informed that the objective of the meeting was to reach a consensus on the mechanics of building an organization to implement a domestic water system. This was identified as a priority need by the people during earlier meetings to create their village natural resource management plan.

The participants passionately debated, among other things, the size of the fee to be collected from members. One of the Council of Elders asked "Why should we pay for water when it has always been a free resource?" The Provincial Development Officer answered: "Why do we complain about a minimal fee of P10 (US$0.20) per month for water services when we do not complain about the much more exorbitant fees of the electric company? What is more basic to farmers, water or electricity?" The community finally agreed on the suggested amount (P10) and also agreed to make use of their indigenous practice of free reciprocal labor (*alluyon*) to build the irrigation and domestic water supply project, while the CHARM project supplied the necessary materials.

This scenario of community consensus building has been taking place to some extent in 82 villages in Abra, Benguet, and Mountain Province in CAR. Out of a total (beneficiary) population of 139,000, 90% are indigenous peoples. These are the Kankanai of Mountain

Province and Northern Benguet, Ibaloi of Southern Benguet, and Tinguian of Abra.

The CHARM project aims to reduce poverty in CAR by increasing annual disposable incomes of smallholder farm families in the target area from US$820 (in 1994) to US$2,170 by 2006. In 1991, 70% of the population had incomes below the poverty threshold, which was about US$1,500 per annum for a family of six. The project goal is to reduce the number of poor families to not more than 25% of the population. The project began in 1997 and will end in March 2003. Total cost is US$41.4 million, obtained through a loan from ADB and the International Fund for Agricultural Development (IFAD).

The components of the project were based on participatory rural appraisals undertaken in 47 villages in 12 municipalities, initially identified as within the project area. The highest ranked needs identified were access infrastructure (roads, vehicles, footbridges, and tramlines), water for irrigation and domestic water use, and market facilities.

PROJECT COMPONENTS

The project components include community mobilization, natural resource management, rural infrastructure development, agricultural support services, and project management coordination.

Under the subcomponent on community mobilization, communities are organized, trained, and mobilized using a community-participatory approach. Detailed household and community socioeconomic profiles are generated. Data from these benchmark studies are included in a village natural resource management plan, which becomes the framework for any development undertaking in the village. Based on this plan, the community prioritizes its needs and the resources to meet these needs.

Activities under the subcomponent on natural resource management include staff training on community mobilization and participatory planning regarding environmental awareness and land tenure instruments available to communities. Also included is capacity support to DENR, the Department of Agrarian Reform, and NCIP to issue appropriate tenurial security instruments, and funding support for DENR's reforestation projects. The rural infrastructure component involves the construction or repair of farm-to-market roads, footbridges, and tramlines; community irrigation; and domestic water supply.

Activities under the component on agricultural support services include financing of market research, the results of which are fed into community planning processes and disseminated to farmers. The Department of Agriculture-CAR agribusiness unit has been expanded by providing services to farmer groups on markets, promotion and linkages, agribusiness feasibility studies, and continuous business management advice. It provides extension services to increase farmer's awareness of available investment options and the implications for resource allocation of various technologies. Funds are also allocated to support an adaptive research program.

The rural financial service aims to strengthen the organizational capability of small farmers through the formation and development of more than 1,530 savings and credit groups over a 4-year period. It provides training to these groups on a monthly basis over a 1-year period.

CONSULTATIVE PROCEDURES

The CHARM project is being implemented through a process-oriented approach. This means that beneficiary participation is encouraged in planning, implementation, monitoring, and evaluation. The lead implementing agency is the Department of Agriculture. The coimplementing agencies are the regional offices of DENR, Department of Agrarian Reform, National Irrigation Authority, local government units, NCIP, and NGOs.

A noteworthy feature is the effort to involve NGOs directly, particularly in the community development subcomponent. An NGO consortium, the SUCCORED, was contracted to coordinate the participatory household and community profiling activities, draw up the village natural resource management plan, and conduct social evaluation of ongoing activities in the communities.

The CHARM project was approved by ADB in 1997, the year before ADB's Policy on Indigenous Peoples was adopted. However, there was an appendix to the CHARM project proposal on a development strategy for ICCs "to ensure that the informed participation of the ICCs in the project is facilitated." It advised that development interventions "should be compatible with the traditional systems and values of the ICCs in CAR." It also stressed that the program will "use the communities' traditional processes for community action."

At this juncture, it would be premature to judge the overall impact of the CHARM project on indigenous communities. In the absence of a more thorough longitudinal study, this report merely outlines the lessons from the CHARM project's experience based on indigenous peoples' perceptions. These lessons are reviewed in the light of ADB's Policy on Indigenous Peoples.

CONTRIBUTION TO LAND TENURE AND ANCESTRAL DOMAIN ISSUES

In general, there is congruence between the CHARM project's package of development interventions and the expressed needs of the indigenous peoples. This was guaranteed by the participatory processes in the project from conceptualization to implementation. These were the rapid rural appraisals conducted during the preparatory phase of the project and several community meetings during the drafting of the village natural resource management plans.

The project is responding to the aspirations of indigenous communities to secure their ancestral lands/domains through efforts to facilitate the issuance of land tenurial instruments. This component has been slow because of the problems in the implementation of IPRA. The project coped with this limitation by facilitating the delineation of ancestral domains and the processing of CADCs.

The coimplementing agencies have been issuing land tenurial instruments within their respective mandates. For example, DENR has been issuing community-based forest resource management agreements. As of 2000, it had accomplished half of its target of 18% coverage. The Department of Agrarian Reform, for its part, has been issuing Certificates of Land Ownership Award. In 2000, it also accomplished about half of its target of 23% coverage. Note that there is no legal conflict between the DENR and Department of Agrarian Reform instruments and IPRA's CADT/ CALT. There is nothing that should prevent indigenous peoples from converting or incorporating their DENR agreements into CADTs, should they choose to do so. They may also convert their landownership certificates into CADTs should they opt for this instrument.

As of 2000, the project had completed 40% of the farm-to-market roads in Abra, Benguet, and Mountain Province. Besides facilitating market access, the infrastructure projects have enabled households to earn cash income by working with small-scale contractors. Income was also generated by the community reforestation project. One poor landless woman from Bucloc earned a gross income of P7,000 (US$140) for 3 months of work in that project. She spent the money on her sons' school needs, a blanket, and some kitchen utensils. However, she wished that she could earn this amount on a more regular basis so she could have some savings.

INVOLVEMENT OF INDIGENOUS PEOPLES/ INSTITUTIONS

Community participation in the planning phase has been successfully accomplished. This is demonstrated by active community participation in composing the village natural resource management plans. Conducted on an annual basis, these plans identify the priority needs and the specific actions to be implemented within the year to help achieve community goals. The goals and activities identified in the plan reflect the felt needs of the community. As a result, the communities have developed a strong sense of ownership over the plans. In addition, in the process of making them, the indigenous peoples' skills in researching, planning, implementing, and evaluating development activities are honed.

However, community involvement in the implementation and in the evaluation phases has not been comprehensively planned or put into place. This explains why, during field interviews, the community expressed some frustration over the delay in implementing their identified priority needs. The reason for this is the decision of the project supervising unit to return the natural resource management plan drafts to the partner SUCCORED for revision. Pending the submission of the revised and final drafts, many of the identified needs are not being addressed. In some instances (e.g., Benguet), the identified needs were realigned by higher levels of decision-making structures from the Municipal Management Group to the Provincial Management Group, to the project supervising group, and the final plan implemented did not reflect the priorities intended by the community.

The project has initiated activities to enhance indigenous institutions. In Abra, where the reciprocal labor system is being enhanced, the project's efforts at facilitating the delineation of ancestral domains, securing tenurial instruments, and establishing interagency cooperation are laudable.

However, there is still no systematic and institutional attempt on the part of the project to enhance (or make use of) viable indigenous institutions, especially on resource management. For example, in Sagada, Mountain Province, traditional social institutions in the management of communal forests persist. People feel free to extract forest resources for their daily needs without charge or fees. This incentive motivates people to protect their claimed ancestral forests from other claimants, fire, and unsustainable use. Consequently, forest protection becomes a collective responsibility of every member of the community. In Abra, a similar practice is *lapat,* a customary system of regulating and managing the use of natural resources.

There are opportunities for incorporating these institutions into the resource management component of the project. The *lapat* system can be transformed into a management board that can organize a group of sheriffs to watch for offenders against forest protection rules. However, there are no such attempts on the part of DENR, the agency charged with the reforestation and agroforestry aspects of the project.

There have been complaints that DENR's reforestation project is not appropriate to the ecosystem of the Cordillera. This sentiment was articulated by Mr. Philip Tinggonong during the project's provincial workshop in June 2001:

The DENR will have to be reeducated (on indigenous resource management) *because they are the villains* (sic) *regarding these things. Look at the kind of species they are planting in the watersheds of the Cordillera. If it is not Gmelina it is mahogany and these are not for watersheds. Instead of focusing on indigenous species to preserve and conserve the diversity they bring in exotic species. According to some foresters, these species are water users instead of water enhancers. When we confronted them they said it is hard to collect the seeds of indigenous species; it is easier to buy the seeds of Gemelina because it is just scattered all along the roads in lowlands.* (Proceedings of the provincial workshop, 4–5 June 2001)

Tinggonong's remark resonates with critiques by national and international indigenous peoples' organizations of commercial reforestation projects being supported by ADB and the World Bank.

LESSONS LEARNED

There is no doubt that the CHARM project has benefited the indigenous peoples in the Cordillera. As mentioned earlier, 90% of the target beneficiaries are indigenous peoples. But do the benefits trickle down to the poorest of the poor indigenous peoples? This is a question that requires more systematic study. For now, there are only initial observations.

In tackling the issue of equitability, it would be good to ponder on the question "If a poor person is landless, will she/he be a beneficiary of an irrigation project?" The impression from the village-level consultations is that, indeed, the CHARM projects have mainly reached people living adjacent to the roads and the relatively well-off, that is, those who have the land and resources (time and money) to attend community consultations.

One controversy involves a case in Benguet, wherein a bid to implement the road construction project was awarded to a private contractor rather than to a community-based group (also cited in a Cordillera Peoples' Alliance letter to ADB, 2001). The head of the rural infrastructure development unit of the CHARM project justified this action:

According to my understanding, if ever the contractor makes some anomalies, you can still trace him because he has an identity. While if you give the money to the beneficiary and if he makes some anomaly you cannot run after him because you may not know him. The money is a loan not a grant. They see to it that the money should be returned. (Proceedings of the provincial workshop, 4-5 June 2001)

He articulates an official development orientation that emphasizes the need to generate rates of return (or return of investment). He also illustrates the tension between economic notions of "efficiency and social development goals of indigenous peoples' participation.

One of the weaknesses of the CHARM project is the seeming contradiction between "output" and "target" orientation in ADB guidelines and the "process-oriented" participatory approach. According to the provincial workshop participants, the former tends to be rigid and time-bound, while the latter needs to be flexible and requires more time to bear fruits. Furthermore, the latter is easily quantifiable, while the former is not. The challenge, therefore, is to reconcile these two approaches.

The other weakness of the project is the inadequacy of local counterparts (local government units and peoples' organizations) both in terms of funds and labor, which contributes to the delay in project implementation. There are also weaknesses in SUCCORED. This NGO is generally weak—it lacks the capacity to carry out the enormous tasks expected of it. This has been attributed to problems internal to the consortium, many of which were already evident at the start of the project. These problems are, to name a few, interorganizational and interpersonal conflicts, lack of experience of some staff, and weak managerial and administrative skills.

The CHARM project's main strength is its participatory approach to development. It has built the capacity and confidence of indigenous peoples (IPOs, NGOs, and local government units) to plan, implement, and evaluate projects. It has also developed the good practice of interagency cooperation and complementation. Another strength is the project's effort to secure tenurial instruments for the indigenous peoples' ancestral lands/domains. It has also responded to the expressed needs of indigenous peoples for greater market access and food sufficiency.

The ADB project in the Cordillera may yet be considered a success story by 2003, if and when the following recommendations, based on the authors' observations, are taken into consideration to make sure that indigenous peoples' culture is consciously considered in policy formulation. The project must

- implement its policy of enhancing and incorporating indigenous structures and processes in the project cycle, and integrate indigenous natural resource management practices with the reforestation program;

- mainstream indigenous peoples' rights and development concepts in all government line agencies and local government units;
- develop the capacity of indigenous people's organizations at the community level to implement the project, with little intervention from the local government unit;
- reinforce/strengthen the capacity of NGO partners to implement the project;
- review its policies and guidelines such that the goal of productivity and sustainability is achieved along with equity and social justice; and
- reconcile the perceived contradiction between its social reforestation project and the ADB Policy on Indigenous Peoples.

11

Conclusions and Recommendations for a Plan of Action

This study affirms that indigenous peoples' indeed have distinctive perspectives on poverty and development. The indigenous resource persons claim that powerlessness, deprivation from access to their land and resources, lack of knowledge (due to lack of education), insufficient income, and alienation from kin/ clan and their culture form the key indicators of poverty. The most frequently mentioned causes of poverty among them are dislocation from ancestral domains and limited or no access to resources in their territories. Displacement is in turn caused by the intrusion of mainstream "development" projects and programs, militarization, and landgrabbing by settlers/ migrants. Official information on ethnicity and development is extremely limited. However, available data reveal that indigenous peoples are not necessarily the "poorest of the poor" in the Philippines. Their regions are relatively wealthy, but extreme inequality, poor infrastructure, and massive exploitation contribute to the worsening poverty situation of these communities.

There exist various opportunities for combating poverty in indigenous peoples' communities. There is in place, a law, the IPRA, that recognizes indigenous peoples' rights and lays down the basic policy framework for indigenous peoples' development. The Supreme Court has recently affirmed the constitutionality of the said edict. In addition, there are several initiatives by international/multilateral agencies, government agencies, and NGOs that attempt to alleviate the plight of indigenous peoples. Although these initiatives are not expressly "poverty reduction" efforts, their projects and activities, when and where properly designed and implemented, do address the main causes of indigenous peoples' poverty. The case study on the ADB/IFAD-supported CHARM project in the Cordillera points out the weaknesses and strengths of antipoverty projects for and by indigenous peoples.

In these poverty reduction efforts, feedback from indigenous peoples indicates the urgency of balancing projects and programs with the basic guiding principles of trust and transparency, nurturing interpersonal relations as a ground for constructive cooperation, and adhering to a nonextractive engagement with stakeholders. Furthermore, these need to be translated into methodology and approaches used in engaging with indigenous communities.

Without these bases for interaction, the enthusiasm for projects and programs may encourage activities conducted in a manner that promotes further impoverishment (e.g., exclusion, powerlessness and dependency, ignorance of indigenous knowledge systems, and loss of affinity with ancestral domains) as defined by the indigenous peoples.

Examples are roads constructed within ancestral domains, employing indigenous peoples as mere laborers; introduction of new technologies without studying the local knowledge systems; making plans for rather than with the indigenous peoples; and application of conflicting tenurial systems within the indigenous peoples' ancestral domains. The experiences narrated by the indigenous peoples call for a restructuring of the articulation of terms such as "participation" (e.g., there is a demand for representation even during the conceptualization stage of the project as well as in monitoring implementation) and "capacity building" (e.g., consultants to become facilitators working closely with and not for the indigenous peoples) within the context of indigenous peoples' own views of poverty reduction and development.

Following is discussion on the formulation of an action plan for indigenous peoples' development, based on the present project's national workshop and the Framework for the NCIP Program of Action (2001).

Detailed recommendations for the action plan are provided in tabular form at the end of the discussion.

The urgency of implementing the IPRA and dealing with the issues raised in the action plan has been highlighted by the reported appearance of a new militant group, the Indigenous Peoples Federal Army.

SECURING INDIGENOUS PEOPLES' ANCESTRAL LAND/ DOMAIN

Access to and control over the ancestral land/ domain is crucial for indigenous peoples' well-being. Moreover, the ancestral domain provides the context and environment that make development possible. Sustainable development for indigenous peoples can only take place by securing the ancestral domain.

To date, most ancestral domains with completed CADCs have not undergone on-ground survey and delineation, and therefore remain vulnerable to various competing claims. The thrust of a development program for indigenous peoples should, therefore, be to identify, delineate, and title ancestral domains. This thrust is expected to meet the objective of recognizing ownership by the indigenous peoples to their ancestral domains. The major programs/projects/activities required to meet this objective include cultural mapping of the communities, survey and delineation of ancestral domains, and issuance of CADTs.

Cultural mapping is important in validating the existence of indigenous peoples' communities—their identities, demographic profiles, geographic distribution, socioeconomic situation, and needs; and their distinct culture, in terms of language, indigenous knowledge, systems, practices, customs, and traditions. Cultural mapping is a prerequisite to any program/project/activity that seeks to direct services to the really deserving communities. As an initial step, there should be an inventory of baseline data on indigenous peoples from government agencies and NGOs; NCIP should provide a specific format for data gathering and should consolidate all collected information.

Another project/activity that is necessary for the issuance of CADTs and CALTs is the identification, validation, survey, mapping, and delineation of the ancestral domain areas. NCIP lacks the capacity to undertake this alone. OPAIPA suggests that

NCIP tap the services of existing experts from the NGO community and academe.

Experts on survey, 3-D mapping, and use of geographic positioning and information system facilities will train a pool of indigenous peoples' leaders and NCIP personnel who will eventually be deployed in mapping and delineation work in their respective areas to fast track the whole process of CADT and CALT issuance. A simplified but effective and efficient procedure in mapping and titling will be developed in collaboration with other government agencies, particularly the Bureau of Lands, the Lands Registration Authority, and NAMRIA.

The titling of ancestral domains includes the systematic validation, certification, and registration of CALTs and CADTs; and the recording and archiving of ancestral domain maps and related documents. One of the foremost tasks is the establishment and adoption of new guidelines for titling ancestral domains, thereby safeguarding the integrity of the process of issuance of indigenous lands and domain titles and the Certificate of FPIC.

MANAGING AND PROTECTING THE ANCESTRAL DOMAIN

One major question that arises is "What becomes of the ancestral domain after it has been awarded to indigenous peoples?" A major objective of development plans for Philippine indigenous peoples is to enhance their capability for sustainable development and protection of their ancestral domains. To achieve this objective, the participants at the project's national workshop proposed a number of activities under the Ancestral Domain Sustainable Development and Protection Plan program. This program should be rooted in indigenous perspectives of development. As mentioned earlier, development for indigenous peoples is a holistic concept that weaves through human, natural, and supernatural realms. Indigenous peoples' notion of well-being is connected with their ancestral domain. Development must also recognize and strengthen their rights to self-determination. It must ensure their participation in all levels and phases of development.

Protection refers to safeguarding the ancestral domains and the conservation, rehabilitation, and protection of the ancestral domain ecology. The IPRA (Section 1 Part IV) defines the role of indigenous peoples in maintaining ecological balance: *Based on their indigenous and traditional practices,* [indigenous peoples communities] *shall formulate and implement their respective systems of protecting and conserving the flora and fauna, watershed areas, sacred places and all other objects of ritual and ecological importance in order to preserve, restore and maintain a balanced ecology within their ancestral domains. To ensure biological diversity, sustainable indigenous agriculture shall be encouraged while the system of mono cropping shall be discouraged.*

NCIP has listed several activities for the management and protection of ancestral domains. Included in the list are formulation of the Ancestral Domain Sustainable Development and Protection Plan, developing indigenous peoples' capacity to manage such plans, community development projects or the provision of social services, scholarship programs for members of their communities, adjudication of conflicts involving indigenous peoples, consensus building among them for the issuance of the Certificate of FPIC, quick response to indigenous peoples in distress, and ancestral domain security and protection. This project's national workshop provided significant points that help clarify the issue of mainstreaming indigenous peoples' concerns in government agencies. The first point is that mainstreaming is not synonymous with assimilation or integration. On the contrary, it should be based on respect for cultural diversity. The second point is that mainstreaming efforts should start with the process of changing the mind-set (world views) of government personnel with regard to indigenous peoples. They should veer away from patronizing and stereotyped ideas about indigenous peoples and instead promote trust and acceptance of indigenous peoples as equal partners in the development process. The third point is that mainstreaming means integrating viable indigenous sociopolitical structures into the mainstream (state) structures and processes.

Drawing from the experience of mainstreaming gender issues in development, the participants in the national workshop proposed a similar approach: government agencies should allocate a certain percentage of their budget for indigenous peoples' development issues and concerns, and create an indigenous peoples' desk (focal point).

To ensure the delivery of basic services to indigenous peoples, NCIP should also coordinate closely with ongoing national initiatives to deliver basic social services to their communities. Examples are CIDSS, described earlier, of the Department of Social Welfare and Development, and KALAHI, a program of the National Anti-Poverty Commission.

CAPACITY BUILDING MAINSTREAMING DELIVERY OF BASIC SERVICES

Access to basic services, livelihood programs, and basic social infrastructure is a priority for these communities. Integral to the ongoing attempts to address indigenous peoples' poverty is the need to mainstream indigenous peoples' programs in government agencies and to coordinate interagency support services for the communities on education, health, and other basic social services. Self-determination is of utmost importance to indigenous peoples. It refers to their right "to decide on their own priorities for development affecting their lives, beliefs, institutions, spiritual well-being and the lands they own, occupy and use" (Rule IV, Part I, Sec. 7, IPRA-IRR). Self-determination includes the exercise of their right to participate in the formulation, implementation, and evaluation of local and national plans, policies, and programs directly affecting them.

Toward this end, NCIP will pursue programs promoting self-governance and support their efforts to organize or strengthen indigenous sociopolitical structures in their communities. Moreover, measures will be undertaken to strengthen structures and systems toward stronger collaboration between indigenous peoples, civil society, and local government units in the development of ancestral domains.

On governance, the challenge is to transform NCIP to become a model of good governance and an instrument of social justice and service to the millions of indigenous Filipinos. To strengthen and revitalize NCIP, there is a need to setting up mechanisms for recognizing genuine indigenous peoples' leaders and instituting policies for meaningful indigenous peoples' participation in government structures.

- cause a radical change in the mind-set of NCIP employees with regard to their understanding of the IPRA, their relations with the community and their attitudes toward public service;

- renew and reenergize its bureaucracy equip it with knowledge, skills, and resources to meet its duties;
- address its "problem of credibility" vis-à-vis indigenous communities; and
- address issues on the legal interpretation of IPRA vis-à-vis other national laws.

NCIP should be instrumental in the empowerment of indigenous peoples. Toward this end, NCIP should help in the institution and capability building of their organizations. The need for sustained strengthening of such organizations within the ancestral domain at the local, regional, and national levels is vital to indigenous peoples' self-determination and to obtain reforms to fully implement the IPRA. Capability building would involve strengthening traditional structures of leadership, governance, conflict resolution, and unity building; and The need for a component on development administration and cooperation emerged during the project's national workshop. This was based on the observation that despite several government initiatives on indigenous peoples' concerns, there is an apparent lack of a cohesive integrated approach to indigenous peoples' development, lack of coordination between and among government agencies with regard to indigenous peoples' development concerns, and no real "presence" yet of NCIP in existing interagency structures and mechanisms.

The workshop clarified that NCIP should play a leading role as far as ancestral domain/land is concerned. Further, NCIP should play a coordinating role in the delivery of social services and livelihood programs, and governance and empowerment.

The objective of the component on development administration and cooperation is to establish more cohesive interagency coordination from the highest executive level down to the village level. The national workshop proposed the crafting of an Executive Order, signed by the President, that would pave the way for the mechanisms for interagency cooperation.

14

Participatory Forest Management in Bangladesh

With the Pace of Population booming, increased energy consumption, over exploitation of the natural resources and rapid depletion of the forest reserves accelerated natural disaster like flood, drought and cyclones in Bangladesh.

Once Bangladesh was famous for its evergreen/ semi evergreen tropical and world famous mangrove forest. But over the years due to over exploitation of forests and its non-participatory management, more than 50% of the forest resources has been depleted. Realising the grim effect of destruction of forests and to repair the lapidated environmental condition, both the government and nongovernment organization have taken up afforestation programme. The NGOs have added a new dimension in the forest management, which has ensured participation of the community people and protection of the vegetation. Although, the government has also adopted participatory forest management but due to bureaucratic attitude easy access of the poor habitants are restricted in many cases. To overcome these situations, the existing government forestry policy, which was formulated in 1994, needs radical modification. There should be room to accommodate the NGOs, grass root organisations and general people in policy formulation, execution and evaluation of the programme.

Introduction

Bangladesh lies in the North-Eastern part of South Asia between 20°34' and 26°38' North in latitude and between 88°1' and 92°41'

East in longitude. The total area or the country is 144,000 sq. km with a population of about 120 million, density of population is 800 person per square kilometer. The most densely populated country in the world, Bangladesh is mainly a floodplain delta, which is formed at the confluence of the Ganges, the Brahmaputra and the Meglina rivers. Natural forest represents only 6 percent of the total land area of the country and is managed and controlled by the government. Village forest which contains annual and perennial trees and still provides a major source of food and income for majority of people, is managed by private individuals.

A rapidly increasing population is placing growing demand on natural resources, especially forest sector is under pressure to become more productive and efficient to keep pace with increasing demand. At present forest in Bangladesh is in unfavourable situation in terms of meeting increasing demand and also not adequate for maintaining ecological balance. This is primarily due to heavy population pressure and limited resource base, secondly, lack of integrated planning for development of multiple resource base with active participation of people resulting in high degree of environmental degradation, as illustrated mostly by deforestation and destruction of Natural resources.

In the phase of rapid depletion of forest aggravated by increasing demand for forest resources, and considering the prevailing socio-economic condition of the country, Government has put emphasis on participatory approach in development of forest resources of the country.

Forest situation in Bangladesh

Bangladesh has lost over 50% of its forest resource over the period of about 25 years. Actual forest coverage is only 6 percent of the total area and the situation is worsening despite of an attempt to preserve it. At approximately 0.02 ha per person of forest, Bangladesh currently has one of the lowest per capita forest ratio in the world.

In Bangladesh, government-owned forest area covers 2.19 million ha, with the remaining 0.27 million ha being privately controlled homestead forests. Of the government owned forest land, 1.49 million ha are national forests under the control of the Department of Forest, with the rest being under control of local governments. Of the state owned forests, over 90% is concentrated in 12 districts in the Eastern and South-Western region of the country. However, due to over exploitation these forests have become seriously degraded.

The natural forests of the country are classified into three categories: 1) Tropical evergreen/ semi-evergreen forest in the eastern districts of Sylhet, Chittagong, Chittagong Hill Tracts, and Cox's Bazaar: 2) Moist/dry deciduous forest also known as Sal forests in the central and the northwest region and 3) Tidal mangrove forest along the coast, known as the sundarban, the largest mangrove ecosystem in the world. These forests are official reserves and placed under the jurisdiction of the Forest Department. Unfortunately, recent inventories indicate a continuing depletion of all major forests.

Forest management in Bangladesh

In Bangladesh management of government forest is the responsibility of the Forest Department under the Ministry of Environment and Forest. In this process the department is managing, protecting, developing the forest resources, forest land and also collecting the revenues. People have never been consulted nor involved in forestry activities. From the management point of view, forest of Bangladesh are being divided into three categories such as:

- State owned forest under the administrative control of Forest Department.
- State owned forest under the administrative control of Ministry of Land through District administration.
- Private village forest managed by private individuals. Forest under Forest Department control and management again divided into three major types viz; (a) Hill Forests; (b) Plain land Sal Forests, (c) Mangrove Forests.

Hill Forests: The tropical evergreen/semi evergreen forest cover as approximately 1.32 million ha of which 0.67 million ha is controlled by the forest department and rest is under the control of hill district council. Clear felling followed by replanting with suitable species (both long and short rotation) is the method of management in hill forest. Because of increased demand for timber and fuel wood and prevailing socio-economic condition of the country this forest has greatly affected and rate of denudation is considerably high. The forest department is mainly confined in raising of single species plantation. Inventory shows that most of these plantations would not give the desirable output. This programme suffers from technical, social and administrative soundness. Another problem is most of the high forest are subjected to shifting cultivation by the hill tribes. The tribes are entitled to shifting cultivation in forest land under administrative control of district administration which has resulted in the total destruction of these tropical evergreen forest. The growing stock has depleted from 23.8 million m^3 in 1964 to less than 20.7 million m^3 in 1998.

Mangrove Forests: Known as Sundarbans, the largest mangrove ecosystem in the world. Sundarban forests are being managed by selection felling method followed by natural regeneration. Beside Sundarbans, plantations are being raised with mangrove species in the newly accreted char land all along the Coast of the Bay of Bengal. Sundarban forest is an official reserve forest, unfortunately recent inventory shows a continuous depletion due to over-cutting, illegal felling. It is estimated that in less then 25 years, the volume of commercial species Sundari, Gewa, has declined by 40 to 50% respectively.

Plain land Sal Forests: Silvicultural system applied for Sal forest was coppice with standard system. In this system matured trees were felled and the areas were protected for coppice regeneration. The typical nature of Sal forest is that this forest is scattered. In the forest areas there are agricultural lands owned by the adjacent people. Frequently these land owners are extending their lands and encroaching to forest and in the process they are destroying the forest and subsequently converting the area to agricultural land. In this process forest lands are being marginalised day by day. FAO estimated that only 36% of the Sal forest cover remained in 1985; more recent estimates that only 10% of the forest cover remains due to over exploitation and illicit felling through there is an official base on logging since 1972. Most of the Sal forest are now substantially degraded and poorly stocked. The situation calls by for involvement of community people in the forest management.

Inventories how that there has been overall depletion in forest resources in all major state owned forest. The growing stock in

Sundarban has been depleted from 20.3 million m^3 in 1960 to 10.9 million m^3 in 1998. In the Hill forest of hill districts, the growing stock has depleted from 23.8 million m^3 in 1964 to less then 20.7 million m^3 in 1998. Over-cutting by timber merchants, increased consumption linked to population growth, shifting cultivation, encroachment, illegal felling and land clearing for agriculture, lack of participatory management have been the principal causes of deforestation and shrinking of forest land in the country.

Since 1960 two major approaches regarding the role of forestry in development have been reflected in the forestry sector of Bangladesh. In the 1960's, Bangladesh as a part of Pakistan and then as an independent nation has followed 'An Industrialisation Approach' consonant with the international conventional wisdom at that time. As a result, Department of Forest raised large-scale Industrial plantation which were seen as conversion of low-yielding natural forest into artificial plantation of species (mostly teak) of great economic importance. This conversion of semi-evergreen and evergreen forest into deciduous teak plantation was largely concentrated in hill forest areas. During the plantation raising local people were not consulted and often they did not drive any benefits from these plantations. The lack of support by the local people/ communities in combination with lack of silvicultural knowledge and lack of proper maintenance contributed to raise low quality plantations and these plantations were also lost due to illegal felling. In the name of plantation the genetic resource of the ever-green/ semi-evergreen forest was lost. Forest Department was considered as revenue earning department. The main activities of Forest Department were concentrated in extraction of trees from the forest and replanting of those felled areas where applicable, Forest Department has not considered the people and their participation in managing forest of the country.

In the 1980s following a change in thinking about the role of forestry in development, and peoples participation in forestry activity was encouraged. People participation with the forestry sector realised the need of people oriented forestry programme to replenish the degraded forest resources of the country. Accordingly, in 1994 Government formulated a forest policy replacing earlier one enunciated in 1979 with a due emphasis to the need for people's participation in forest management.

PARTICIPATORY FOREST MANAGEMENT APPROACH IN BANGLADESH

Past activities

Forest extension activities were formally launched in the country in the year of 1964 with the establishment of two forest extension divisions at Dhaka and Rajshahi and later two divisions at Comilla and Jessore. It was really a very small programme and the activities were confined only to establish nursery in the districts headquarter and raised seedling and sell the same to individuals and organizations. The location of this programme was so urbanized and limited that it only partially served the needs of the effluent town dwellers only.

Betagi-pomora comunity forestry project

The first community forestry programme in the country, started at Betagi and Pomora mouza (village) under the district of Chittagong in the year of 1979 with the personnal initiative of Prof. A. Alim, renowned forester and Prof. Dr. Mohammed Yunus, founder of Gramen Bank. Initially the project covered 160 ha of Government denuded hilly land at Betagi and with 83 landless participants from adjacent community and subsequently extend over another 205 ha of Government owned denuded hilly land at Pomora with another batch of 243 landless (families) participants. Under this programme each landless participant was provided with 1.62 ha of land for growing tree and hoticultural crops with technical and financial assistance from the Forest Department.

This community programme has given the landless an identity of their own and a sense of direction in life. But this model has not been replicated in the other areas due to lack of initiative of the Forest Department as well as the Government.

Rehabilitation of Jhumia families (shifting cultivator families)

Another project was undertaken by the Forest Department in the Hill tract areas to establish plantation through rehabilitation of Jhumia families in 1980. Main objectives of the programme were (i) to rehabilitate tribal families in the Unclassed State Forest(USF) lands along with rehabilitation of denuded USP land;

(ii) to introduce a sustainable agroforestry production system; (iii) to improve the socio-economic condition of the tribal people and (iv) to motivate tribal people in development of forestry. Under this programme each family was allocated 2.02 ha of USF land for growing agricultural crops (over 1.20 ha), raising plantation (0.80 ha) and for house construction (0.20 ha). The rehabilitated families were given land use rights and were allowed to enjoy 100% benefits accrued to those lands.

The participants were given input support for growing agriculture, horticulture and forestry crops and cash support for house construction. This programme continues for quite a long period of time but could not sustain mainly because of nomadic character of the tribal groups. Another reason of failure was that the families were rehabilitated in clustered villages without considering their cultural and religious values. Thus in most of the cases it was found that the families have left the area. A parallel programme was also initiated by the Chittagong Hill Tract Development Board in which Forest Department was responsible for implementation of afforestation component where Cittagong Hill Tract Development Board was responsible for the rehabilitation component. This program was also not found so much responsive to hilly people except for some plantation establishment.

Development of community forests project

The activities of the first phase of this project began in 1981 and were completed in 1987 in seven greater districts of the North-Western zone of the country. The main components of the project were:

- Strip plantations along roads and highways, railways, canal sides, district and Union Parishad roads, totalling about 4,000 km.
- Fuelwood plantation on 4800 ha of depleted Government land on participatory concept.
- Agroforestry demonstration farms over 120 ha also with participatory concept.
- Replenishment of depleted homestead wood lots in 4,650 villages.
- Training of Forest Department Personnel and Village leaders.

Development of forest extension services (1980-1987)

Development of Forest Extension Services (Phase II) began in 1980 with the Government funding and subsequently amalgamated in some areas (i.e. North-North West district) with Asian Development Bank funded Community Forestry Project. The main activities under this programme were:

- afforestation in some 3100 villages.
- roadside tree planting along 3600 km of primary highways and roads and about 600 km of Union Parishad roads.
- production of 49 million seedlings for distribution.

Thana afforestation and nursery development project

This project is a follow-up of Development of Community Forestry Project and Forest Extension Project and has been designed primarily to: (i) increase the production of biomass fuels and (ii) enhance the institutional capability of FD and local administration in implementing a self-sustaining nationwide social forestry programme.

In order to increase the production of biomass fuel and to arrest the depletion of tree resources, the project envisaged to develop tree resources base through planting of depleted sal forest as well as brining all suitable and available land in the rural areas under tree cover with active participation of the rural poor of the locality.

Originally the project was to be implemented by the Forest Department and former Thana Parishad during the period of 1987 to 1994. But in 1992 Government decided that the all project activities were to be implemented by Forest department alone. The major components of the project were:

1. Establishment of plantation over 20,225 ha depleted Sal forest areas.
2. Development of agroforestry over 4,200 ha in the Sal forest lands.
3. Raising strip plantation on 17,272 km along Road and highway, Railways, Embankment and Feeder Roads.
4. Raising 1,282 ha plantation in the land outside the BWDB.
5. Planting 7.017 million seedlings at the premises of different education, religious and social institutions

6. Establishment of 345 nurseries at Thana headquarters.
7. Raising of 10.618 million seedling for distribution to public.
8. Training of some 76,000 people of different levels.

Here this may be mentioned that at the last stage of the project implementation, the Government has found that this was quite impossible to protect the strip plantation and also impossible to trained 76,000 people by the Forest Department alone. The Government invited NGOs to participate in this programme for successful implementation. PROSHIKA, POUSH, GRAMMEN BANK and other NGOs came forward to help the Government for successful completion of the project; NGOs employed their group members to protect the strip plantation and ADAB came forward to train people at different levels with the help of its member organisations.

The above plantation activities were carried out with the direct participation of the local people with the help of the NGOs by executing benefit sharing agreement.

Coastal greenbelt project

Another project financed by Asian Development Bank is under implementation in the Coastal region of Bangladesh. The main objective of the project is to create a vegetative belt all along the coast to save the lives and properties of the people living in the coastal areas from devastated cyclone and tidal surges which occur very frequently in those areas. All of the activities of this project are also being carried out following participatory approach. In this project also the participants have been selected among the poor people living in the adjacent areas by involving NGO and a pre-designed benefit sharing agreements also being executed with the participants to protect their rights over plantations and to ensure benefit expected to be received out of the plantation.

Agroforestry research project

Pilot Agroforestry Research and Demonstration was implemented by the FD in the Sal forest areas. The project had been developed precisely to design/develop agroforestry modules which is environmentally feasible, socio-economi-cally acceptable enhance tree and crop production at the same time to uplift the socio-economic condition of the participants. The project aimed at using 120 ha of encroached Sal Forest land of Dhaka, Mymenshing and Tangail Forest Division to develop suitable participatory plantation models.

Food assisted social forestry programme

The World Food Programme assisted the Government to develop Social Forestry as a national programme and the Government incorporated WFP assisted social forestry programme in its annual development plan from 1998. Poverty alleviation, economic rehabilitation of rural poor especially the destitute women of the society by engaging them in forestry activities, social uplift of rural poor and environmental improvement are the main objectives of this project. Historically this programme was conceived in the country since 1989 on pilot basis allocating in kind resources (Wheat) to a limited number of NGOs for raising strip plantation along roads, embankments, Highways etc. in rural areas following the participatory mechanism. In implementing this programme FD was involved later on to provide technical guidance to the NGOs and other GOB agencies. At present probably this is the largest Participatory Forestry Programme in Bangladesh. From 1990, 100 NGOs are involved in this programme and at present about 60 NGOs are continuing with the programme. Commencing from 1990 up to 1998 about 31 million trees were planted involving 0.062 million people directly and 0.62 million people indirectly. The programme has created employment to the tune of 68 million man days. This programme is being implemented by the NGOs through contractual benefit sharing among participating poor men & women 60%, NGOs 10%, the rest land owners.

NGO PARTICIPATION IN THE PARTICIPATORY FOREST MANAGEMENT

In Bangladesh the history of NGO involvement in the field of development is not very old. After liberation, NGO started their activities through relief and rehabilitation of the war victims. During mid-seventies, NGO switched over to the socio-economic development of the rural poor, and at present there are thousands of NGOs most of whose mandate is to organise rural poor and provide awareness, education, skill training and various support services including credit to enhance participation of landless poor in the development process towards self reliance. On the basis of the networking throughout the country the NGO can be classified into two levels; (i) local and (ii) National. At present more then 100 NGOs both local and National are implementing social forestry programme in Bangladesh. It is not possible to enlist activities of all the NGOs involved in the Social forestry programme in the country. The participatory forestry activities of some of the NGOs are highlighted here who are playing pioneer role in this field.

BRAC:The Bangladesh Rural Advancement Committee (BRAC) has been established in 1972 and this is the largest NGO in Bangladesh. BRAC has six categories of projects/ programmes in broad sense among which Social Forestry falls within rural development programme. The Social/ Participatory forestry has three components; (a) Nursery establishment; (b) Plantation and (c) Establishment of agroforestry. They claim that they have assisted their women members in raising more then 225 homestead nurseries and 100 large nurseries with a combined production capacity of a million seedlings of fruits and forest seedlings. They have established 200 mulberry nurseries with the production capacity of 2 million seedlings. Social afforestation programme of BRAC is WFP assisted which has been commenced from 1989 and till now it is going on. Most of the plantation under this programme has been established along the strips. Up to this time about 33.72 million seedling have been planted over an area of 33,700 km strips along road, railway and embankment. This programme involves about 670,000 participants of which 80% are women.

PROSHIKA:Proshika-A center for human development is one of the largest NGO in Bangladesh. The Social Forestry Programme of Proshika is a systematic intervention effort to enhance afforestation in the country and to make a case that the poor are the best managers and protectors of forest resources if they are granted usufruct rights on these resources. Proshika has introduced its group members in social forestry activities and provided them with credit and technical support, which contributed significantly to their self-sufficiency. The main components of social forestry programme of Proshika are; i) Homestead plantation; ii) Strip and block plantation; iii) Natural Forest protection, and iv) Nursery establishment. Proshika has planted 71 million seedlings which covered along 8,887 km strips, 37,662 areas of block plantation which included natural Sal forest protection throughout the country. One of the most significant contributor of Proshika to the development areana is the introduction of the concept of participatory forest management for natural forest protection. Proshika has successfully involved the forest dwellers in the Sal forest areas of Kaliakoir, Mirzapur, Shakhipur and Shreepur thanas under district of Tangail and Gazipur for the protection of coppice Sal forest by involving group members of Proshika. It has already been proved that when poor people surviving on the forest resources are organised, trained and granted usufruct rights, they present on enormous human potential needed for afforestation and forest protection.

RDRS:The Rangpur-Dinajpur Rural Services operating in 28 thanas of greater Rangpur and Dinajpur districts covering 28 thanas. It is the largest International Integrated Rural Development NGOs operating in Northern Bangladesh for more than two decades. Its entry point in forestry was through road side plantation in 1977. Initially, seedling were protected with bamboo cages. Situation has been changed a lot nowadays and protection of seedlings with bamboo cage has become a part of history. Besides strip plantation, they also extend their tree plantation programme in homestead, institutional grounds as well as raising of local nurseries. With the assistance of WFP they have planted about 10.66 million trees under their participatory afforestation programme.

TMSS:It stands for Thangamara Mohila Sabuj Sangha. It is an NGO exclusively meant for women. Although, it was initiated in 1976, its presence was visible only since 1965. This NGO believes in the concept of simple living and high thinking. TMSS was also involved in the social forestry programme particularly in the Northern districts of Bangladesh. The organisation has been implementing both road side and farm forestry with assistance from the WFP and Swiss Development Corporation (SDC) respectively.

POUSH:Another NGO has been engaged in planting in the private lands also by persuading the owner farmers in Baroibari, Kaliakoir thana with the food aid from WFP. Its activities in the forestry field are limited to strip plantations and it plans to expand its programme extensively. Target groups of POUSH are mostly landless destitute, widow and divorced women. POUSH also happens to be the first of its kind to get involved in participatory forestry in the hill district of Bandarban

PROBLEMS AND PROSPECTS OF PARTICIPATORY FOREST MANAGEMENT

Considering the demand and supply situation for forest products to meet the economic and environmental needs, no one can deny the need of people's participation in forestry. Probably there is no second answer except participatory forestry in developing, managing, and protecting the country's forest land and the forest resource. But there are numbers of issues remain unresolved. As a technical department, Forest Department is playing pioneer role in implementing and popularising Participatory Forestry in the country. Up to this time Forest Department is managed by the professional foresters who have educational background only in managing traditional forests and who do not consider people as development partners. Participatory Forestry, if we recollect the Chinese proverb, needs mental development managers towards the people.

Realisation has started among the planners, policy makers, administrators and Senior managers to involve people in forestry development activities. But up to this time Government has failed to adopt real Participatory Forestry programme to address the basic need of the peoples.

Mobilisation of the people in participatory forestry programme is another bottleneck of the Forest Department who has not had the machinery to reach the community people. NGOs who work at the grassroots level have developed their own expert to mobilise people and ensure their participation in any development programmes as partner. So NGO should be involved in the implementation of the participatory forestry programme where Forest Department should confine their activities only in technical aspect.

In the context of Bangladesh, the scarcity of land is a most vital problem. On the other hand, Forest Department controlling over 16% of the total land area of the country is still hesitant to allow Participatory Forestry in reserved forest areas. According to FD, it should be confined only in public and private lands beyond reserved forest areas through these are devoid of trees. The

Participatory Forestry is being practiced in marginal lands which are under administration control of other Government departments. Recently, due to pressure from planners and donor communities, Forest Department has allowed to practice participatory forestry in Sal Forest areas. Tenure of the contract was found as a bottleneck for implementing Participatory Forestry. Forest Department allowed rights of participants over these lands for a period of seven years, but there was a strong desire, that this tenure should more and at least for rotation period, so that participants can manage and protect trees till harvesting.

A negative attitude was also observed among the Foresters to involve women in forestry activities. They viewed that activities of women should be confined in the areas where there is a locality apprehending the social problems. But in participatory forestry both men and women should be treated equally.

Recommendations

- The forest policy of 1994 needs a fundamental change to make participatory forestry approach as a core concept for social fencing against forest destruction and for poverty alleviation through income generation. An independent Forest Policy needs to be formulated to promote Participatory Forestry in the country.
- For implementing Participatory Forestry programme, Land as an input is to be ensured with authority. Preset land lease system is not responsive nor effective for practicing Participatory Forestry in the country. Tenurial rights of land instead of land use right is to be given to the participants.
- To make effective participation of the people on a sustained basis sufficient motivation is required and this responsibility must be given to the NGOs:

Involvement of the groups in planning and decision making is to be ensured through proper policy directions.

- Provision should be made for giving subsistence to the participants so that they can survive and keep confined their activities in the programme.
- A National forum where there will be representation all from Government politicians, NGOs and private sector.

They will act as a coordination body to coordinate among all the participants like land owning agency, NGO, Forest Department and the participants.

- NGO should act as a catalyst and they should not be treated as a competitor of Forest Department.
- Process of recruiting NGOs by inviting tender for implementing any Participatory Forestry programme of the Government must be avoided. ADAB may be given the responsibility to identify the NGOs for a particular programme implementation.
- Major reforestration and afforestation programmes with community participation in a profit sharing basis and improvement in technology and management of plantation is required.
- Institutional capacity, policy and legal framework of forestry management should be strengthened.
- Forestry professionals should be trained in environmentally sound forestry.
- Implementation strategies for community management of forestry resources should be developed.

13

Forests and Forestry Research in India

INTRODUCTION

In view of the deteriorating forest resources and their importance to the national economy and environment, the Government has been emphasizing for the sustainable development of forest resources, as well as conservation of ecosystems. The first national level planning exercise in the forestry sector took place two decades ago when the National Commission on Agriculture (NCA) studied the situation in the country and made recommendations. But actions did not have the desired effect on the development of forestry. The Government revised the National Forest Policy in 1988 and emphasised on conservation and environmental stability. Later on, it decided to undertake the preparation of a National Forestry Action Programme (NFAP) with the aim of addressing the issues of the forestry sector in line with the National Forest Policy (NFP 1988).

Our forests have been transformed drastically and radically, and find it is hard to any undistributed forests as created by Mother Nature. In the present system, forests are unable to provide an intrinsic protection, the production of multiple products or environmental and human functions. Drastic changes are, therefore, warranted to avert an ecological disaster and the national catastrophe that would inevitably follow. Clear cutting has turned out to be a dangerous prescription unsuitable for tropical forests, not even for the European forests where it has originated. The tropical forests are inadequately understood at best and misunderstood at worst; the traditional management practices are the best suited. Forestry research in India is one of the oldest in the tropical world. It was not long ago looked upon as a model for other countries in the region. Guidelines for the preparation of working plans for a wide variety of high and coppice forest, yield regulations, preparation of yield and volume tables, thinning technologies and the like were among the outstanding achievements. Monumental works such as *Silviculture of Indian Trees* by Troup, *Forest Insect* by Benson, *Forest Floras* by Kanji Lal, Talbot Brandis, Mooney and others, speak for the high standard of work of Indian forestry research, which reflects the high degree of dedication.

The present situation

(i) In India, forests meet nearly 40% of the energy needs of the country, of which more than 80% is utilized in the rural areas, and about 30% of the fodder needs of the cattle population. Forest products also play a very important role in the rural and tribal economy, as many of the non-wood forest products (NWFPs) provide sustenance to the rural poor and its accounts for some 70% of forest products and for some 40% of state forest revenues (Ahmed 1997). For landless families and marginal farmers, forest – related activities often represent the primary source of income.

(ii) India's forests are unable to meet its many needs. Ganguli (2000) reported a current round wood supply gap (excluding fuel wood) of some 26 million m^3: The gap is expected to remain in future also.

(iii) It is estimated that about 270 million tonnes of fuel wood, 280 million tonnes of fodder, over 12 million m^3 of timber and countless non-wood forest products are removed from the forest annually. At a conservative level of pricing (Rs. 500 per tonne of fuel/fodder) the value of these commodities will amount to over Rs. 3,00,000 million per annum.

According to FSI (1999), in about 53% of natural forests, natural regeneration is either absent or inadequate. This situation is because of overgrazing in forests beyond their carrying capacity, which has troublesome implications for future growth. A recent study in Kerala has estimated the current average increments, in the natural evergreen and moist deciduous forests as 1.0 and 0.5 m^3 ha^{-1} yr^{-1} respectively. Yet another report suggests that the average productivity of Sal forests of India is at the level of 54.7 million m^3 ha^{-1}. In fact figures are often quoted which would give Indian forest an annual productivity of 0.5 m^3 ha^{-1}. If theoretical potential productivity could be achieved, this would increase to about 525 million $m^3 ha^{-1} yr^{-1}$. These estimates are indicative of a lack of proper management. It is to be understood that the result of management lapses is not limited to growth and yield of timber, but to the survival of critical ecosystems.

Factors influencing productivity

For a given species or forest type, productivity is the function of habitat factors i.e. soil, nutrient status, rainfall and silvicultural management practices. Productivity with respect to forests can be expressed in term of mean annual increment (MAI). A natural forest, with a continuous cycle of growth grows on the products of its own decay (humus) through nutrient recycling. Productivity is influenced by special factors like removal or retention of litter, status of ground vegetation, density of stock on which further growth is to take place and protection form growth retarding agents. It is understood that the MAI of natural forests tend to be lower than those of plantation forests raised with high level of inputs. However, the potential biological productivity of the natural forests of India is definitely higher than the figures mentioned earlier. Using Patersons index and based on rainfall, soil condition and length of growth period, the estimated productivity of these forests would range from 1.35 m^3 ha^{-1} yr^{-1} in the dry regions of India to 7.66 m^3 ha^{-1} yr^{-1} in the moist and wet regions with most other parts of the country producing around 4 to 5 m^3 ha^{-1} yr^{-1}.

Forestry resources of India

Most of the forest ecosystems of India are fragile and reeling under an acute form of degradation. Each ecosystem has specific problems, which make the associated research dimensions vast, complex and demanding. The following factors are attributed to the present state of forest biocoenosis:

(i) Loss of forest land to agriculture, industries and human settlements.

(ii) Loss of forest due to multipurpose projects, construction of roads, erection of transmission lines, quarrying, slashing for shifting cultivation and clearing for encroachments for minerals etc.

(iii) Degradation due to illicit felling, lopping for fodder and fuel wood, overgrazing, forest litter removal, forest fires, over felling, etc.

(iv) Human and cattle population explosion around forest land.

The forests of India have a growing stock of 4,741 million m^3 with an annual increment of 87.62 million m^3. Since 78% of the forest area is subjected to grazing, coupled with heavy removal of forest products and 51% subjected to occasional fire, the productivity of these forests is low. The performance of plantations, in terms of survival, growth and yield is also poor because of inadequate site selection, improper site species matching, poor quality planting stock, lack of post establishment protection and lack of timely silvicultural operations. The average MAI for plantations of the fast growing species under social forestry programme is only 4 m^3 ha^{-1} yr^{-1} against the expected 10 m^3 ha^{-1} yr^{-1}. According to a study of IIPA (1989), the average survival percentage in five Indian states was 66 per cent. Lack of plantation inventory, absence of yield improvement plans and rotation cycle and lack of post establishment management systems were some of the reasons. Yields obtained are far short of the potential. Research support is absolutely essential for realizing their true potential.

ROLE OF FORESTRY SECTOR IN ECONOMY

Contribution to GDP

India, in spite of having 2.5% of the world's geographic area and 1.8% of the world's forests, sustains 16% of the planet's human population and 18% of its livestock population. Forestry contribution is 1.7% of nation's GDP. This does not take into account unrecorded withdrawals (NWFP, fuel, wood, fodder etc.). Moreover, the environmental benefits of forests also remain to be quantified and calculated.

The forests contribute 1.7% to the GDP of the country (NFAP 1999 a & b). Due to problems associated with the valuation of forests and services, unrecorded removals, illegal harvesting, etc. the exactness of the contribution has not been established. A large part of the forest production consisting of fuel, fodder, medicine and food are removed without payment and without any record by the rural and tribal people. According to Ahmed (1997), the total annual value of India's harvest of all forest produce is estimated to be Rs. 300,000 millions (compared to the investment of Rs. 8000 in the sector). The low estimate of contribution to the GDP resulted in low priority for forestry investments in five-year plans. Efforts are needed for monitoring the services provided by the forests so as to appreciate their contribution to human well being. Over 50% of the revenue earned by the forest departments comes from NWFPs. Their growth is generally 40% higher than timber (MOEF 2000). Nearly 350 million people living in and around forests in India depend on NWFPs for their sustenance and supplemental income which is worth Rs.400 billion annually (Tewari 1994). Studies in Orissa, Madhya Pradesh, Himachal Pradesh and Bihar have indicated that over 80% of forest dwellers depend entirely on NWFPs. Similarly 17% landless depend on daily wages related to the collection of NWFPs. 39% people are, however, involved in NWFPs collection as a subsidiary occupation. It has been estimated that many village communities derive as much as 17-35% of their annual household income from the sale of NWFPs (Tewari 1994). NWFPs provide 50% of the income to about 30% rural people. The average income realized through the sale of NWFPs by households in the state of Madhya Pradesh constituted 34 to 55 percent of their total income. As per estimates made in West Bengal, an average return of Rs 2270 ha yr^{-1} is obtained from NWFPs, which is 25% more than the polewood harvest, which fetches Rs 16,000 per ha after 10 years. There is, thus sufficient evidence to believe that the collection of NWFPs is a crucial part of the population's life support system, especially of the tribals.

Employment generation

Of the total wage employment in the forestry sector, NWFPs account for more than 70% of the opportunities for self-employment for the forest dwellers as farm mechanization has not developed well in India. According to an ILO estimate, one hectare of forest plantation creates nearly 630 mandays, from the raising of nurseries to the harvesting stage. 70% of the budget allocated to plantations or afforestation is spent on providing direct wages to the workers and only 30% goes towards purchase of seeds, planting materials, equipment etc. It would not be out of place to mention that 50% of the workforces on forest plantations are women and tribal. Rural women use 70-80% of the mandays in collection of NWFPs, fuel and fodder. Activities related to NWFPs provide employment during slack periods and a buffer against risk and household emergencies. In the remotest areas, sometimes the forest is the only source of employment and income. Research is needed to evolve forest based entrepreneurial endeavours to produce multiplier effects through the forward and backward linkages.

Markets

The dependence of the producer on intermediaries and his limited access to markets has a direct effect on prices. The price of a product whether sold to consumers directly or through intermediaries, has no bearing on the expenditure incurred on the labour, inputs and transportation. Under direct sales, the localized activity for localized markets creates a supply position in excess of local demand. Traders control the market and dictate the prices during the season and in the off-season. The sale of produce during the flush season and in the off-season is different. In the case of sale through intermediaries, the producers have absolutely no control over the prices. Studies show that the poor primary producer's income always remains low. The need for market related studies has always been felt, and includes research on the market information system and the scope of value addition like bioprospecting, Intellectual Property Right (IPR) etc.

Growing pressure on forests

The present ecological conflicts have created many economic compulsions and sociological stresses due to changing consumption patterns, scarce availability of land and other natural resources.

Out of the total requirement for wood, 70% is for fuel wood and 30% for timber. Thus, forests have at least 5 times more pressure than what they can withstand. This is in addition to the 30% contribution to the fodder requirement of the country in the form of 178 million tonnes of green fodder and 145 million tonnes of dry fodder.

Supply and demand scenario of forest raw materials

The current demand of timber is around 30 million m^3, out of which 8.3 million m^3 is needed for paper, pulp and panel products and 15.4 million m^3 for saw milling i.e. housing, packaging, furniture, etc. The total timber requirement is estimated to grow to a level of 60 million m^3 during the next five years (20.5 million m^3 for paper, pulp and panel products and 27 million m^3 for saw milling). Therefore, during 1994-1995, nearly US$ 600 million worth of wood and wood products were imported. Similarly, the current requirement of fuel wood is around 280 million tonnes, which is likely to rise to 356 million tonnes. As such, there is an ever enlarging gap in the supply against the current sustainable yield of only 12 million m^3 of timber and 21 million tonnes of fuel wood.

Despite the low consumption and production, shortage has been the history of the India's forestry industry sector. India is, therefore, supplementing its wood with bamboo and agriculture wastes. The trend, due to poor transport economics force Mills to be small and within the transport radius. The situation of Indian paper mills, depending on the sugar mills is within the 100 km radius, for cool-fired boilers for their own heat needs. To spare the bagasse for sale as pulping raw material there is severe pollution in pulping agriculture radius and poor quality paper relative to wood fiber material. This trend is likely to continue if there is no reform in the forest policy linking biophysical, demand, research and marketing.

MEDICINAL PLANTS

Herbal medicines in use today are derived from nearly 8,200 species of medicinal plants. In developing countries, more than 80% of the population depends on traditional plant based medicine, and even in the USA, 25% of the prescription drugs are still based on phytochemi-cals. NWFPs offer an excellent potential international trade. Ahmed (1997) reported 8 categories of NWFP, some are important from the point of tribal indigenous groups such as West Bengal where tribal groups collect 27 commercial products, 39 plant food, 47 medicines. NWFPs account for 70% India's exports and the demand for phytochemicals is expected to increase in future as a new frontier for trade and sustainable commercialization of biological resources at international level with secured intellectual property right.

The manufacturers and the practitioners of traditional medicines depend solely on the procurement of MFP plants from the wild, as cultivation of most of these plants barring the spices has not gained momentum. The methodology of extraction of phytochemicals from the forest products has gained attention of researchers. The value additions may be done through percolation methods using extraction of plants, spray drying, distillation, effleurage or cold fat extraction, maceration or hot fat extraction and solvent extraction. Oleo-resins are obtained conventionally by extraction of the powdered spice with solvents, but notable differences in the yield and in aroma flavour and pungency of the final product can be achieved by varying the geographical type of the spice, the age at which it is harvested, the choice of solvents and the method of extraction. To make this possible a determined investment in technological capacity building, biophysical research and development, market research is crucial.

The major task before the country, therefore, is to rehabilitate the degraded forests and to enhance the area under forest/tree cover to 33% of total area as envisaged in the National Forest Policy (1988). Despite continuous efforts the desired results could not be achieved because of the ever-increasing demand for forest products and the limited funds available (less than 1% of the total Plan allocation). To reverse the process of degradation and to achieve the goal of the National Forestry Policy (1988), the government formulated the National Forestry Action Programme (NFAP), a com.

Research priorities

In the policy document (GOI 1988) a number of priority areas of research and development, needing special attention have been identified as a research strategy to achieve the objectives of the National Forest Policy. These are:

- Increasing the productivity of wood and other forest produce per unit of area per unit of time by the application of modern scientific and technological methods.
- Revegetation of barren/marginal/ waste/ mined lands and watershed areas.

- Effective conservation and management of existing forest resources (mainly natural forest ecosystem).
- Research related to social forestry for rural/ tribal development.
- Development of substitutes to replace wood and wood products; and
- Research related to wildlife and management of national parks and sanctuaries.
- In the light of the sector review (GOI), the strategic issues for forestry research are identified as follows:
- Increasing forests and tree productivity.
- Enhancing peoples participation in forest management.
- Saving the biological diversity: in-and-ex-situ conservation.
- Rehabilitating degraded lands.
- Strengthening institutional and human capacity; and
- Improving forestry research policy and setting priorities.

Goal and objectives

- Increasing productivity from less than 1 m^3 ha^{-1} yr^{-1} to 5 m^3 ha^{-1} yr^{-1} expanding forest/tree cover to achieve national goal of 33% of geographic area.
- Extending protected area network for biodiversity conservation.
- Planting of fuel wood species on non-forest wastelands.
- Producing pulp wood and other industrial wood through agro-forestry.
- Encouraging and strengthening of JFM.
- Developing NWFPs and generating rural employment through value addition.
- Promoting development, harvesting, processing and marketing of medicinal plants for income generation.
- Strengthening infrastructure for forest inventory, research and development.
- Augmenting research capabilities of state forest departments.
- Building the capacity and training of personnel engaged in forestry through well conceived HRD planning.

- Building the capacity for proper monitoring and evaluation of forestry projects at the state and central levels.
- Rationalizing investment for the sustainable development of forests.

In the early part of this century, forestry research concentrated on botany, taxonomy, phenology, silviculture, logging and wood utilization. However, in the present day context the scope of forestry research covers not only biological and technological aspects (forestry, forest products, conservation and wildlife), but also the wider spectrum of economic and sociological researches leading to policy formulations.

Conclusions

In developing countries like India, an increase in the population of humans and cattle results in an increase in demand for forest produce, where there is scope for cultivation of forestry species in wasteland and other forest land because of the drastic decrease in per capita land availability. Forest produce has to be increased from the land already available and stress sites have to be cultivated to meet the demand. Under these circumstances the research should be linked with the ongoing programmes on tree management. Improvement studies should be given priority in order to achieve meaningful results. This approach is sadly lacking in many developing countries including India, where research and development in forest and improvements yet to show meaningful results.

14

Saving the Life of Tribals

Improving the quality of life of Scheduled Castes, Scheduled Tribes, Minorities and people living in the geographically isolated areas viz. Char, Tea gardens and Border Areas and measures to improve the Government Delivery System therein.

To undertake study and to make specific recommendations as contemplated in ToR 5, the most important requirement is status papers with reliable up-to-date data and information in respect of people like Scheduled Tribes, Scheduled Castes, Tea Tribes, Minorities as well as people living in the Char and Border Areas. But unfortunately such data and information are not available easily. For implementation of result oriented, need based development programmes for the benefit of the aforesaid categories of people, the Commission recommends that people/area specific techno-economic study with analysis should be taken up with scientifically designed parameters. Assam Tribal Research Institute may take up such study in respect of SC, ST and OBC. Directorate of Economics and Statistics may cover Minorities, Char Areas and Border Areas.

On the basis of data and information whatever made available to the Commission, the Commission has formulated the following broad recommendations for improvement of delivery system as well as implementation of development programmes with the objective of economic empowerment and also general development of these weaker sections.

WELFARE OF ST, SC, TEA TRIBES AND CHAR AREA PEOPLE

In this context, it may perhaps be clarified that the Scheduled Tribes, discussed here are ST Plains excluding Hill Tribes of NC

Hills and Karbi Anglong Districts, whose welfare is looked after by the respective Autonomous District Councils

The WPT & BC Department looks after the Welfare of ST (P), SC, OBC and Tea Garden Tribes people. The department carries the welfare activities through the following Directorates:

1. Director, Welfare of PT & BC
2. Director, Welfare of SC
3. Director, Welfare of Tea Tribes

The Directorate of WPT & BC implements welfare schemes through the District and Sub-Divisional offices. The plains districts of Assam are divided into 19 Integrated Tribal Development Projects (ITDP) for implementation of the Tribal Sub Plan (TSP). For each ITDP there is a Project Director who is an ACS (Senior Grade) officer. There is a Project Implementation Committee (PIC), with a non-official chairperson, to approve beneficiaries and to review implementation of the TSP Schemes. For implementation of schemes outside the TSP, there are 43 Sub-Divisional Welfare Offices (SDWO). In each SDWO, there is a Sub-Divisional Welfare Board, with a non-official Chairperson, to approve beneficiaries, to select areas and institutions outside TSP areas and to review implementation of schemes. There is also a Sub-Divisional OBC Welfare Board for implementation of the welfare schemes for OBC people. The Directorate implements schemes for promotion of Education, Income Generation (FOIGS) and Infrastructure Development (roads, bridges, school/college buildings, community halls, and drinking water).

The Directorate for Welfare of SC implements welfare schemes through the SDWOs. The beneficiaries of the schemes are selected by the Sub-Divisional SC Development Boards, which have non-officials as Chairpersons. It implements scholarship schemes through the Project Director, ITDP. The Directorate also provides grants to the Autonomous Councils for Rabhas, Misings and Tiwas for the benefit of the poor SC people in the council areas. Two more Autonomous Councils one .for Deuries and the other for Sonowal Kacharis have been formed recently.

The Directorate for Welfare of Tea Garden Tribes (including ex-Tea Garden people) implements schemes for the welfare of tea-tribes population. For economic development of the people it provides grants-in-aid under "Family Oriented Income Generating Schemes" to BPL persons. For promotion of education, scholarships are awarded and grants-in-aid are provided. It also supports welfare works and cultural activities by NGOs.

ASSAM TRIBAL DEVELOPMENT AUTHORITY (ATDA)

The Assam Tribal Development Authority was set up in 1983 under an Act of the State Government. It was set up for socio-economic development of tribal people and to formulate schemes within the framework of State Plans. ATDA has notified the ITDPs and SDWOs as agents. With funds received under Special Central Assistance (SCA) from the Government, it implements schemes for income generation and infrastructural development. It also implements a Vocational Training Scheme under the Central sector.

Corporations

The following corporations have been established for the welfare of ST (P), SC and OBC.

1. Assam Plains Tribal Development Corporation
2. Assam State Development Corporation for SC
3. Assam State Development Corporation for OBC

The Corporations mainly look after socio-economic development of BPL people.

Char Areas: The whole tract of river Brahmaputra covering about 800 km in Assam is having 3608 sq km char areas (4.6% of the state area)

The Char areas are distributed in 23 Sub-Divisions falling under 14 districts, with 2251 villages in 299 Gaon Panchayats and 59 Development Blocks. The total population of the char areas is 24.90 lakh with 12.72 lakh males and 12.18 lakh females, i.e. 958 females per 1000 males, and the total number of families is 4.35 lakh of which 2.95 lakh (68%) are below the poverty line. The Char areas include Majuli with a population of 2.15 lakh. Density of population per square kilometrer in Char Areas is much higher than average density in the State (690/sq. km. against State average of 340/sq. km.).

The literacy in the char areas is 19.31 per cent. The literacy for the areas in Jorhat district (where Majuli is situated) is 60.55 per cent and for the other districts it varies from a minimum of 12.34 per cent for areas in Darrang district to the maximum of 18.50% for the areas in both Marigaon and Lakhimpur districts.

Protection of international boundary is the responsibility of the Central Government, whereas protection of inter-state boundary is the responsibility of the concerned State Governments. Inter-state boundary is constitutionally well-marked and well-defined, but some states have disputed the notified boundary leading even to border clashes resulting in loss of lives and properties, necessitating intervention by the Central Government. The two Boundary Commissions were constituted one for demarcating boundary between Assam and Nagaland under Shri K.V.K. Sundaram in 1971 and another to demarcate the boundary between Assam and Meghalaya under Justice Y.V. Chandrachud, former Chief Justice of India in 1985. Both the learned Commissions after mature deliberations and detailed study of historical records upheld Assam's claims, which was rejected by the State Governments of Nagaland and Meghalaya respectively. At present the Hon'ble Supreme Court is hearing the boundary dispute between Assam and Nagaland.

Boundary dispute with Arunachal Pradesh, Mizoram and Meghalaya could not be settled in spite of several rounds of discussion at the highest level. There was agreement to have joint survey of boundary, but this also could not proceed due to last minute objections of the neighbouring states. Tension continues to prevail in some border areas, though the concerned state governments have agreed to maintain status-quo.

In respect of international boundary, there is clear guidelines of the Government of India for development of border areas contiguous to the international boundary. The Central Government have treated all blocks situated at the international boundary as border blocks which are considered as spatial unit for implementation of all development schemes sanctioned under Border Areas Development Programme started in 1993-94, initially for infrastructural development in the bordering areas of Indo-Bangladesh border, which was later extended in 1999-2000 to Indo-Bhutan border. 15 blocks in Assam-Bangladesh border and 10 blocks in Assam-Bhutan border have been identified as border blocks and all schemes drawn up under Border Area Development

Programmes are implemented through the respective blocks. The Border Area Development Programme is 100% centrally funded under Special Central Assistance Schemes allotted as grant on the basis of (i) length of the international border, (ii) population of border blocks and (iii) area of border blocks. In fact the Planning Commission has formulated specific guidelines for Border Area Development Programmes with the objective of meeting the special needs of the people living in remote and inaccessible areas situated near international borders.

The schemes cover not only development of infrastructures like road, bridges but also cover schemes for economic benefit like agriculture, fishery, horticulture, healthcare, education, vocational and technical education, social forestry, rural sanitation, drinking water supply, training for skill up-gradation, employment generation programme etc.

The scheme also envisages involvement of village panchayat at all levels of decision making for selection of schemes, implementation and monitoring.

In respect of border areas of inter-state boundary, there is no clear demarcation of boundary of the border areas. As a result even basic statistics like number of villages falling in border areas, population, number of schools, dispensaries etc are not available with the Directorate of Border Areas. Therefore government delivery system has not even remotely touched the villages lying close to inter-state border, who are constantly living in fear of attack from the neighbouring states. No development work has taken place, no welfare schemes even reached the people living in the remote villages. Even basic amenities of life like drinking water supply, schools, dispensaries are absent in those area.

Whatever fund is allocated, they are used for constructions of rural roads, bridges, BoPs etc. Other sectors such as health, education, sanitation etc are completely neglected. As a result, people are not attracted to reside in the border areas. Taking advantage of sparse population, large number of undesirable elements are reported to have occupied vast tract of border areas which would pose a threat to the territorial integrity of the state.

Due to remoteness of the area and lack of basic amenities, lower level government functionaries such as school teachers, nurses, village-level workers, Extension officer etc. are reluctant to stay in the border areas. Their visit to those areas is also less frequent. As a result people in the border areas are deprived of the benefits of most of the welfare schemes launched by the State Government with much fanfare. Non-involvement of village–level institutions at any stage of the decision making process is another reason for the failure of the government delivery system.

General Recommendations

1. For cost-effective efficient delivery of government services/ benefits to ST, SC, Tea Tribes, Minorities, residents of Char Areas, Border Areas, the number of delivery agencies should be restricted to one for each category.
2. For ST, at present we have five statutory Autonomous Councils for Bodo, Mising, Tiwa, Rabha-Hasong, Deuri and Sonowal Kacharis. The councils have covered all the core areas of respective tribes. Delivery of government services/benefits to STs residing in the Autonomous Council areas of Assam should be entrusted to these councils.
3. STs outside Autonomous Councils could be covered by ITDPs (which is a Central Government pattern). After the formation of Autonomous Councils as mentioned above, the ITDPs would be having spare time and infrastructure, which should be utilized by the State Government for managing delivery services for the benefit of SC and OBC. This could be done with minor structural adjustment without any financial implication.
4. Assam Tribal Development Authority was constituted in 1983 as an umbrella organization to look after the development of ST population of the state living in plains. But after the formation of so many Autonomous Councils to look after different tribes, the role of ATDA has become redundant and this organization may perhaps be abolished.
5. As a measure of economy, instead of appointing a separate Principal Secretary for the smaller Autonomous Council, the Project Director of ITDP could be made Ex-officio Principal Secretary. This would result in considerable saving of the meagre resources now allotted to these councils. For this purpose headquarters of the concerned Project Directors should be located accordingly.

6. Directorates of SC, ST etc. should confine their activities to supervision, guidance evaluation and monitoring. The Directorates need not be involved in delivery services; as has been done now. At present, the same quantum of fund allocated for development of STs has been transacted by a number of delivering agencies, thereby enhancing the cost of such delivery which should be avoided. Government guidelines should stipulate the maximum percentage which should be spent as overhead cost. It should be examined if all the aforesaid Directorates could be merged into a single Directorate of Social Empowerment.
7. For welfare of people working in Tea gardens statutory obligation of the management to provide amenities to tea garden workers should be strictly enforced.
8. For development of Char Areas as well as economic empowerment of BPL families, a road map should be carefully prepared by expert/experts. The Master Plan prepared by the Minority Welfare and Development Department for Micro-Credit Extension in Char Areas (prepared with the help of Geography Department of GU) is a good step in this direction. Similarly, compilation of statistics published by Directorate of Char Areas Development under the caption "Socio Economic Survey Report 2003-2004" is praise worthy, and these figures could provide basis for undertaking techno-economic study of the issues pertaining to Char Area development. But the road map suggested by us should be a realistic one covering all aspects of Char development, prepared on the basis of techno-economic study and analysis, providing methodology for implementation of programmes on the basis of the road map.

8:1 The Mobility of Char dwellers is greatly restricted by physical barriers, since there is no transport linkage not only between Char areas and mainland, but also within the char areas. Immediate steps should be taken to improve the communication facilities as otherwise all efforts to improve the socio economic condition of char dwellers would not be successful.

9: For economic empowerment of BPL Char families Pilot Scheme for economic empowerment through individual skill development backed by need and skill based micro-credit should be tried through Self Help Groups.

10. If the Pilot Schemes mentioned above succeed, this may be the pattern for economic empowerment not only for Char BPL families but for other weaker sections like ST, SC, Tea Tribes, Border people etc. Implementation of individual Family Oriented Schemes would be hardly of any help to tackle this gigantic problem of economic empowerment.
11. For providing elementary coverage to BPL families, relating to health & hygiene, literacy etc. help of reliable NGOs may be utilized who should function through committees of actual beneficiaries at local level.
12. Following the Government of India guidelines, the State Government may identify the border development blocks and specify gaon panchayats located in the inter-state border areas for implementation of special development programmes for the benefit of people living in border areas.
13. After identification of the border block, and gaon panhchayat detail survey be made to assess the needs of the people and the critical gaps in physical and social infrastructure so that a perspective plan for development of border areas can be prepared and implemented.
14. While preparing schemes, Planning Commission guidelines (Annexure) be followed so that besides infrastructural development schemes embracing education, health, agriculture, social sector can also be prepared.
15. Gaon panchayat and Anchalik panchayat be involved in selection of schemes, implementation and monitoring.
16. State Government may consider earmarking one percent of fund from all development departments for Border Area development programmes for which a non lapsable fund could be created. This would increase the availability of fund for development of border areas.
17. Government officers working in border areas be given remote area allowance of 15% of his basic salary as an incentive to work in the interior areas.

18. The Directorate of Border Areas be upgraded to a Commiserate with a Commissioner of Border Areas whose main functions would be-
    - I. Overall administrative charge of all border blocks both at the international and inter-state boundary.
    - II. To prepare schemes for border areas development both at the international boundary and inter-state boundary.
    - III. To liaise with his counterpart in the neighbouring states in the case of border dispute
    - IV. To work in close coordination with IGP (Border) to maintain territorial integrity of the inter-state border and also with Divisional Commissioners as suggested by the Commission while recommending decentralization of powers up to Divisional level vis-a-vis strengthening the role and powers of Divisional Commission (Para 14, Functions of Divisional Commissioner)
    - V. To maintain all historical records of the border, so that in case of any legal dispute these can be readily made available.
    - VI. Any of the work that may be allotted by the government relating to the border.
    - VII. People affected by flood, erosion or displaced otherwise may be encouraged to settle in border areas for which exemption of land revenue for the first ten years may be granted as an incentive.
    - VIII. Social forestry may undertake massive programmes in the border blocks for afforestation.
    - IX. The State Government should set up government offices, schools, police out posts, BoPs, dispensary etc. in the border blocks.
    - X. The State Government should compel the field level officers and staff specifically engaged for work in border areas to stay in their respective areas of work.

The Commission is confident that efficient and effective delivery system for the benefit of the people covered under ToR 5 could be set up on the basis of the recommendations enumerated above.

15

Planning Commission (MLP Division)

BORDER AREA DEVELOPMENT PROGRAMME – REVISED GUIDELINES (APRIL, 2003)

The Border Area Development Programme (BADP) was started during the Seventh Plan with the twin objectives of balanced development of sensitive border areas in the Western region through adequate provision of infrastructural facilities and promotion of a sense of security amongst the local population. The programme was revamped in thee Eighth Plan (1993-94), and extended to States which have an international border with Bangladesh. The nature of the programme was changed from a schematic programme with emphasis on education to a State level Programme with emphasis on balanced development of border areas. During the Ninth Plan, the programme has been further extended to States which border Myanmar, China, Bhutan and Nepal and the programme at present covers all the seventeen States which share international borders.

Coverage

Border Area Development Programme covers seventeen States namely Arunachal Pradesh, Assam, Bihar, Gujrat, Himachal Pradesh, Jammu & Kashmir, Manipur, Meghalaya, Mizoram, Nagaland, Punjab, Rajasthan, Sikkim, Tripura, Uttar Pradesh, Uttaranchal and West Bengal.

Objectives

The main objective of the programme will be to meet the

special developmental needs of the people living in remote and inaccessible areas situated near international border.

Funding

The Border Area Development Programme will continue to be a 100% Centrally funded Area Programme. Funds will be provided to the States as Special Central Assistance for execution of approved schemes on a 100% grant basis and allocated amongst the seventeen beneficiary States on the basis of (i) length of international border, (ii) population of border blocks and (iii) area of border blocks. Each of these criteria will be given equal weightage. The border block will be the spatial unit for the programme and all schemes would have to be implemented within the border blocks only.

Eligible Schemes

Schemes should be designed to take care of the special problems faced by people living in the border blocks, particularly in the rural areas. Schemes for rural and remote areas along the international border are to be given preference over schemes for urban areas. A long term Perspective Plan for each Border block should be prepared by the State Government keeping in view the objectives of overall balanced development of the region. The State Government may undertake a study of remote villages in the border blocks in order assess the needs of the people and the critical gaps in physical and social infrastructure. Schemes should be drawn up based on this assessment. Grass root institutions such as PRIs/District Councils/Traditional Councils should be involved in identification of the priority areas to the maximum extent possible.

Since the people living in the border areas must have a direct say in the selection of schemes, village level institutions such as Gram Sabhas should be involved in the decision making process. Appropriate modalities may be worked out by the State Government to ensure greater participation of the people of the border areas in the selection of schemes under the programme. Schemes which address problems such as inadequacies relating to provision of essential needs, strengthening of the social infrastructure, filling up of critical gaps in the road network etc.

may be taken up under the programme. Emphasis must be laid on schemes for employment generation, production oriented activities and schemes which provide for critical inputs in the social sector.

Security related schemes can also be taken up. However, expenditure on such schemes should not exceed 7.5% of the total allocation in a particular year. Construction of housing for crucial functionaries such as teachers, doctors, nurses etc. may be taken up in border blocks under the programme along with construction of small culverts, bridges, bridle paths, feeder roads, etc. However, expenditure on schemes including provision of basic amenities such as supply of drinking water, approach roads, etc., for the border outposts, administrative buildings and road & bridge construction taken together should not be more than 60 percent of the total allocation in any particular year. In addition, up to 15 percent of the total allocation could be used for maintenance of the assets already created under the programme. It may be ensured that no single sector gets a disproportionately large share of the total allocation. The funds under BADP are to be used for schemes in the identified border blocks only. Funds under BADP are additive to normal Central Assistance and are allocated for addressing the special problems faced by the people of border areas. Hence, these should not be used to replace normal State flows.

Executing Agencies

To provide flexibility, schemes under the programme can be executed by any of the following agencies:

(i) State Government

(ii) Central Government

(iii) Central Para-Military Organization located in the State; and

(iv) Voluntary Agencies

(v) Panchayati Raj Institutions/District Councils/Traditional Councils.

Due emphasis should be given to effective involvement of local people/ democratic institutions / voluntary agencies in order to inspire mutual trust and confidence between the Government and the people.

Screening Committee

Subject to such general or special directions as may be given by the Empowered Committee, individual schemes located in each State will be approved by a Screening Committee chaired by the Chief Secretary of the State. The Screening Committee will also include representatives of the Ministry of Home Affairs and the Department for Development of the North Eastern Region (DONER), not below the rank of Joint Secretary to the Government of India, representative(s) at similar level of the major para-military organization(s), SSB operating on the State Border, State Planning Secretary, Secretary, Home Department of the State or Secretary, Border Area Development Department (as convener) and the concerned State Plan Adviser / Adviser (MLP) of the Planning Commission. The State Chief Secretary may co-opt members on the Screening Committee depending on the local requirements.

The Screening Committee will have complete freedom to execute the schemes through any of the five agencies mentioned in para VI above. The Screening Committee shall meet at least once in a year but positively in February to finalize the schemes for the following year and assess the progress of schemes under the programme.

Release of funds, monitoring and review

Before the commencement of a financial year Planning Commission will inform the States of the funds available for the State during the next year under the programme. A list of schemes proposed to be executed within the ceiling communicated, will have to be sent to the Planning Commission for release of funds to the State. For the North Eastern States, the list of approved schemes will have to be sent to the Planning Commission through the Department for Development of the North Eastern Region. Funds will be released in two installments; the first instalment will be equal to two-thirds of the annual allocation, while the second instalment will be the remaining one-third. The first release will be made by the month of April after the receipt of the list of schemes, duly approved by the Screening Committee along with expenditure incurred till the last quarter of the previous year. The Screening Committee must approve the schemes to be taken up in February of the previous year assuming the level of funding prevailing during that year. The State Governments which fail to have the list of schemes approved by 31st March, will stand to forfeit their first instalment. The first instalment will not be released if the unspent balance exceeds the annual allocation for the previous year. The second release will be made after October depending on physical and financial progress but not later than February after adjusting unspent balance, if any of the previous year.

State Governments are required to have a separate budget head for the programme. The State Governments will ensure that Display Boards are placed at every block Head Quarter indicating the allocation of funds under the Border Area Development Programme. Display Boards may also be placed in front of the assets created with BADP funds stating "Funded from Special Central Assistance under the Border Area Development Programme."

Periodical monitoring of schemes in the Planning Commission will be done by the concerned Member/State Plan Adviser/Adviser (MLP). A review of the programme will be made at least once a year by the Empowered Committee.

Quarterly progress reports (proforma at Annexure 2) should be submitted scheme-wise to Planning Commission (through DONER for the States of North East) giving actual physical and financial achievements and not in percentage terms. The quarterly progress reports should be sent as soon as possible so as to enable Planning Commission to recommend the release of Special Central Assistance. The utilization certificates should be sent by 31st January following the year of release.

BORDER AREA DEVELOPMENT PROGRAMME: ILLUSTRATIVE LIST OF SCHEMES

A. Education

a) Primary school buildings (additional rooms)

b) Middle school buildings (additional rooms)

c) Development of play fields

d) Construction of hostels/dormitories

e) Buying of school dresses/books

f) Adult education

g) Public libraries and reading rooms

h) Development of human resource through vocational and technical education
i) TV/dish antennas
j) Books/journals

B. Health:

a) Provision of medical equipments of basic/elementary type, X-Ray, ECG machines etc. can also be purchased.
b) Provision of equipment for Dental Clinic
c) First-aid kit for midwives
d) Construction of public toilets and bathroom
e) Setting up of mobile dispensaries in rural areas by Government/ Panchayati Raj Institutions
f) Health Awareness Programme
g) Eye camps
h) RCH programme
i) Blood Banks
j) Control of Malaria, Filaria, Leprosy, AIDS etc.

C. Agriculture and allied sectors:

1. Animal Husbandry
2. Pisciculture
3. Sericulture
4. Farm forestry, horticulture, pastures, parks and gardens in government and community lands or other surrendered lands.
5. Desilting of ponds in villages, towns and cities
6. Construction of public irrigation(minor work) and public drainage facilities
7. Construction of common gobar gas plants, non-conventional energy systems/devices for community use and related activities
8. Construction of irrigation embankments, or lift irrigation or water table recharging facilities
9. Minor irrigation works
10. Anti-erosion works (only construction of small bunds etc.)

11. Water conservation programmes.
12. Social Forestry, JFM
13. Use of improved seeds, fertilizers and improved technology
14. Veterinary aid Centres, artificial insemination Centres and breeding Centres.

D. Infrastructure:

1. Rural Sanitation
2. Solar Street Lights
3. Solar household Lights
4. Electrification
5. Strengthening of existing roads (only link road)
6. Construction and strengthening of kutcha roads, part road, approach roads, link roads.
7. Drinking water facility – construction of tube wells, water tanks
8. Repair and maintenance of existing roads
9. Public Distribution System and its improvement (in terms of infrastructure)
10. Industries – Handloom, tiny units, handicraft, furniture making
11. Construction of culverts/bridges and roads
12. Drains and gutters
13. Footpaths, pathways and footbridges
14. Provisions of civic amenities like electricity, water, pathways, public toilets etc. in slum areas of cities, towns and villages and in SC/ ST habitations.
15. Provision of common work sheds in slums and for artisans.
16. Bus sheds/stops for public transport passengers.
17. Repair and maintenance works of any type other than special repairs for restoration/ up gradation of any asset.
18. Construction of buildings for local bodies for recognized District or State Sports Associations and for cultural and sport activities or for hospitals

(provision of multi-gym facilities in gymnastic centres, sports association, physical education training institutions etc.)

E. Social Sector:

1. Construction of community centres
2. Rural Sanitation
3. Anganwadies, Crèches
4. Cultural Centres/Community Halls
5. Construction of common shelters for the old and handicapped.
6. Training for youth for self employment and skill up gradation.
7. Setting up of small industries using local inputs.

16

Eco-Tourism for So-Called Tribal Development: A Critical Assessment from an Anthropological Perspective

INTRODUCTION

Over the years Tourism has become an integral part of economic, social, psychological and physical development. It is considered as an important source of economy especially in terms of its contribution towards foreign exchange earnings, generation of national income and to promote local employment particularly in remote and backward areas.

Unlike other developing countries India has also entered the tourism industry since independence. It contributes only 0.3 per cent to World tourism. Considering India's wealth of natural resources and rich cultural heritage, tourism can emerge as an important instrument for economic development and employment generation. According to the report of World Travel and Tourism Council (WTTC), India could generate 25 million additional jobs in Tourism sector by 2010. After five decades of tourism development India's tourism earnings is US $1.8 billion while tiny Singapore's is US $ 3.4 billion and Thailand's US $6.8 billion. Comparing the earnings from tourism sector the policy makers decided to boost the industry, which is also considered as "passport to development".

Today the whole world, particularly the developing countries, faces a near-crisis situation, both economic and environmental.

Policy makers find it difficult to formulate programmers that would work under the present situation of escalating population on the one hand and diminishing resources on the other. The environmental decadence inevitably weakens economy, which in turn leads to social integration.

There is no denying that the earth's environment has been severely compromised in the last 200 years since the beginning of the Industrial Revolution. Ironically, this era of mass production and amassed wealth can account for much of the poverty in the world today. Agro-based industry have displaced family farms; indigenous and communal lands are transferred or sold to corporations and other foreigners and rural residents migrate to urban centers-about 160,000 people everyday–where they face crowd-ed and unhealthy conditions. Concern is extremely high among people living in developed nations over protecting, cleaning and healing our sick environment, but in order to solve the environmental problems of the world, we will have to solve the inequities in all areas of human life (Hargrove, 2002).

Gradually the issue of environmental care has emerged as one of the most important debates in developing countries. In particular, there has been an increased interest in tourism as a potential vehicle for development that is both economically beneficial and environmentally sustainable and this is encapsulated in the buzz word "Eco tourism".

The main objective of eco tourism promoters are to preserve the natural environment and to promote local employment. When preservation is combined successfully with development-the preserve and use mandate-sustainable development is achieved. Keeping this in view, there has been a rapid growth of eco tourism activities in tribal pockets of India as well as in Orissa. Efforts are being taken to use eco-tourism potential for keeping the tribals in promoting their eco-friendly economic pursuits.

Scholars like Mckean (1982), Volkman (1990) Picard (1993) and McArthy (1994) reported the impacts of tourism on indigeneous peoples. In some areas, economic and cultural benefits to the host community were identified (Jordan, 1980; Silver, 1992) including enhancement and revitali-zation of local cultures (Jafari, 1992; Bleasdale and Tapsell, 1996). The cultural revitalization argument is based on the premise that societies find niches through tourism that help build group solidarity, pride in traditions and strengthened identity (Rabibhadana, 1992). This in turn benefits tourists, tourist promoters and national culture (Klieger, 1990). Dogan (1989) has pointed out that impacts both positive and negative have varied between localities and with type of tourism. The reactions of indigenous hosts have been diverse, ranging from an active resistance to passive adoption of western culture. At times, local culture may be treated as a "commodity" as influential western societies interact with less affluent hosts (Rabibhadana, 1992). Alternatively, interactions may be detached primarily economic, rather than social (Esman, 1984). In other circumstances, visitor influx, seasonality and opportunistic investment in real estate that radically affects local land values may indirectly influence host communities (Craik, 1991).

Smith (1989) argued that cultural impact is minimal where the flow of visitors is small and sporadic, while increasing tourist numbers may compromise the cultural heritage of host communities. "To preserve the cultural resources and enhance the economic base, local communities must decide whether to control or restrict tourism to preserve their cultural integrity or encourage tourism as a desirable economic goal and restructure their culture to absorb it" (Smith, 1989; Rabibhadana, 1992).

In the context of sustainable tribal development in a globalised world, cultural dimension can not be neglected (Upali, 1994), rather one has to consider endogenity (Mangaraj, 2000) at one end and the outside world at the other in order to maintain the pace of development with a changing environment.

Further, the International Labour Organisation's (ILO) approach in the area of Indigenous and Tribal people in its Convention No 169 clearly spelt out .... "Participation of indigenous and tribal peoples in the process that affects them and the need to take into account their cultures, traditions and institutions in the adoption or implementation of programs, activities or legislation that affects them". An important aim of the convention is to set up the conditions for self-management so that indigenous and tribal peoples can gain greater recognition of their distinct cultures, traditions and customs, as well as gain more control over their own economic social and cultural development (ILO Geneva, July 2001).

Article 7 of ILO Convention No 169 emphasizes that indigenous and tribal peoples have the right to decide their own priorities for the process of development as it affects their beliefs, customs and lands, into exercise control to the extent possible, over their own economic, social and cultural developments.

In this regard it is very important for the outside agencies to understand the cultural dynamics of the people before introducing eco tourism to the region in the name of development. From this point of view *Panchsheel* (five principles for tribal development) approach for the tribals pronounced by Pandit Nehru should be taken into consideration where he laid down "These people should be allowed to develop along the lines of their own genius. We should avoid imposing anything on them. We should try to encourage their own way of life, their own traditional arts and culture".

Against this backdrop this paper critically examines the concept of eco tourism in the context of environmental conservation and economic benefits. This paper also tries to analyze the on going processes of eco tourism and its linkage with tribal economy and cultural practices. The paper also focuses some of the impacts of eco tourism on tribal livelihood. Finally the paper suggests some specific strategies for promotion of eco tourism for the tribal people.

ECOTOURISM: CONCEPT AND POLICY

With the growing city culture, the civilized man is slowly moving away from the natural environment and therefore there is an increasing demand for ecotourism' all over the world. It is an industry that touches almost every other, embodying a synthesis of goods and services, materials and markets, employment and education, politics and ideals. At its best, it empowers women, local communities, diverse cultures and travellers, supports partnerships between governments, NGOs, private sector, funding agencies, and blends the high technology of transcontinental flight and internet communications with ancient dances and sacred sites. At its worst, it robs people of low dignity, culture and control over their lives (Tchamie, 1994; Mc Laren, 1999, Maikhuri et al., 2000).

International tourism arrivals have climbed steadily from 420 million arrivals in 1989 to over 664 million arrivals ten years later in 1999, equaling more than 10 per cent of the global population (Ceballos-Lascurian, 1991; Epler Wood, 2002). In 2001, tourism worldwide recorded 693 million international tourist arrivals; producing $463 billion in international tourism receipts (WTO, 2002).

Amongst all international visitors who arrive in India, nature tourist comprise 40-60 per cent. As the fastest growing industry in the world (WTTC, 1992), tourism is looked upon by many governments of emerging countries to develop their natural resources and preserve them at the same time.

The concept to involve local people as beneficiaries and stake holders in the eco tourism process was voiced by the World Bank in 1986 as it promoted "rural development investment that provide farmers and villagers in the vicinity of wild land management areas with an alternative to further encroachment" (Honey, 1999: 16). Ecotourism has finally found its niche – helping the world's rural and underdeveloped peoples to enter into the mainstream market economy.

Keeping in view the global importance, the UN General Assembly in Decemember 1998 declared the year 2002 the International Year of Ecotourism (IYE). It clearly stated that the activities related to the Year should be taken within the broader framework of sustainable development of tourism, with four main criteria.

- Sustainable use of biodiversity and natural resources.
- Impact minimization, especially in terms of change and energy consumption:
- Empowerment and fully informed participation of local stake holders, particularly local communities and indigenous people.
- Awareness raising and environmental education of travelers and hosts (Hillel, 2001)

Ecotourism in these contexts joined the ranks of other sustainable industries which were called to meet global challenges and affect change. The UN had the foresight and courage to place ecotourism on a larger backdrop with global climate change and energy issues and to promote participation of local communities to ensure equity in decision making.

DEFINITION OF ECOTOURISM

Debate over the definition of the new term "ecotourism" began in the mid 1980's and still continues to distract the field today. Many scholars insist to move beyond mere definitions and concentrate on action. Numerous guidelines on how to accomplish ecotourism have transformed from a product into a philosophy (WTO and UNEP, 1992; Manidis Roberts Consultants, 1997; Austrilan Heritage Commission, 2000; Ceballos-Lascurain, 2001).

Ecotourism continually evolving as evaluations of successes and failures are incorporated into the knowledge base of worldwide ecotourism practices. Studies by Epler Wood (2002) reveal that "Like all other sustainable tourism, ecotourism is a dynamic field, with new techniques and approaches.......every year" (Epler Wood, 2002: 7-8).

The Ecotourism Society (TES) defined Eco-tourism as a "purposeful travel to natural history of the environment, taking care not to alter the integrity of the ecosystem, while producing economic opportunities that make the conservation of natural resources beneficial to local people. The International Ecotourism Society (TIES) defined ecotourism, "responsible travel to natural areas that conserve the environment and improves the welfare and to sustain the well being of local people" (Wester, 1993; Epler Wood, 2002).

"Ecotourism is environmentally responsible travel and visitation to relatively undisturbed natural areas, in order to enjoy and appreciate nature (and any accompanying cultural features-both past and present) that promotes conservation; has low negative visitor impact; and provides for beneficially active socio-economic involvement of local populations".

Center of Ecotourism (CES) defined "Ecotourism as an enlightening, particularly travel experience to environments, both natural and cultural, that ensures the sustainable use, at an appropriate level of environmental resources and, whilst producing viable economic opportunities for the tourism industry and host communities making the use of these resources through conservation beneficial to all tourism role players".

Further UNEP focused on these aspects:

- Positive contribution to the conservation of sensitive

ecosystems and protected areas, through financial and political support.

- Active participation from and economic benefits to, local communities and indigenous people.
- Environmental education of hosts, professionals and guests (Hillel, 2001).

The concept of ecotourism stands alive and well after being tested in 2002 at two major world fora, the World Ecotourism Summit (WES) and the World Summit on Sustainable Development (WSSD), and numerous preparatory conferences. Ecotourism has become a major thrust of the common wealth government's tourist's strategies (DOT CWITH) 1992, 1994. Ecotourism mainly emphasizes "a set of principles and how to put them into practice" (Honey, 1999: 21). It espouses the same principles of sustainable tourism, but simply on a smaller localized human scale, while remaining committed to the larger ecological scale.

In the International Year of Eco tourism 2002 (IYE), proclaimed by United Nations (UN) eco tourism was both hailed for preserving dwindling cultures and biodiversity and blamed for degrading both environment and human dignity.

The Draft National Environment Policy (NEP) 2004 (Government of India), (A policy for promoting eco tourism and private Investment) appears to be a comprehensive document covering a wide range of issues. The policy content and approach take their cue from the myriad International agreement India is a party to, and does not in anyway stem from a genuine interest to protect the interests of the poorest sections of Indian Society who are most dependent on the environment for their livelihood and survival.

The draft policy openly plays to the demands of the World Bank and other multilateral bodies such as the WTO and is clearly aimed at privatizing the environment and natural resources. Emphasis has been given for public-private partnership. The polluter will pay concept, commodifications and commercialization of biodiversity and culture, the concern for "eco tourism" as a means of earning state revenue.

These are the driving engines of this policy and clearly concern the peoples and their livelihood. The symbiotic relationship and socio-cultural constraints are not a priority.

The policy will only hasten the all out sale and auctioning of India's natural resources and in doing so systematically marginalize those who are most dependent on those sources-the tribals, the dalits, the pastoralists, other small and marginal farmers-particularly women from these communities.

The deep concern about the aspects of this policy (NEP, 2004): under the Principles 4 (V-A) tribals engaged in *shifting cultivations would be told they have to pay for the perceived damage to the environment and the production of "smoke" and "Carbon dioxide". Farmers would be told that a tax is being levied on them for the "*CO_2 *and Methane emissions" produced by dung of their animals and their burning paddy fields. Woman who collects fuel would be told she has to "pay" for having access to the forests.*

Polices have been formulated for the overall development of a society when poor sections of that society don't avail the fruits of development, then policy for whom?

TRIBAL ECO SYSTEM: INTERLINKED ECONOMIC AND CULTURAL PRACTICES

From time immemorial tribal communities all over the world live close to nature. They have in course of time made a long adjustment to the decay of nature, and derived a traditional and primitive way of living. Hence a distinctive culture gets established obviously owing to their own peculiar ways in order to meet their basic needs. The totality of these interactions within them is known as ecological system or ecosystem.

We have empirical ethnographic data outlining the nature of relationship and interactions the tribals have with various elements of creation. An over acknowledgement of the "gifts" received from nature finds its expressions in their worship of its various manifestations-plants, animals, hills, rivers and many other objects. In closer analysis of the tribal world view, it came to limelight that objects of nature are not seen as an inanimate entities by them, but persons, as fellow human beings. This concern institutionally finds its expressions in "totemism" a practice that refers to a cultural phenomenon in which human groups and individuals occupy defined positions in a network of relationship with the objects of eco system. They have a strong belief system, which states that this Universe is the creation of a supreme power meant for the benefit of all his creations. Each individual must, therefore learn to enjoy its benefits by forming a part of the system in close relation with other species. Let not any one species encroach upon the others rights. Thus they have a lot of respect for nature and its objects.

As Prof Radhamohan an eminent environmentalist rightly points out, "The relation between the tribals and forests is like that of fish and water". Forests are not only one of the major sources of their subsistence, but are significantly related to their religion and mythology for e.g. the tribal people of Orissa worship nature and believe that nature is to be kept satisfied if it is so to provide all their needs. Most tribes believed that "Sun God" is the creator and master of the Universe and call it by many names. The Juanga and Bhuiyans call it "Dharam Devata" the Kolha and Sanatal "Sing Bonga". The worship of the earth is commonly called "Basumata" by Santals, Bhuiyan and Juanga "Dharani Panu" by Kondhs 'Basuki Thakurani' by Kolha. The worship of the earth Goddess acquires special significance since every cultivation is started by rituals of worship, for a good harvest.

The Tribals not only worship the forests, but revel in religious ceremonies and festivals connected with it. Bhinjals and Parajas call their forest god " Danger Devata" Bondas. "Uga" and "Remngbori", Kolha "Bura Bonga", Kondhs "Laipenu" and so on. Considering nature as their creator, sustainer and provider, the tribals have imbibed a deep love for nature that is primeval and instinctive. The Kondhs of Ganjam claim descent from a woman, whose body parts are supposed to be made of "Bel" Fruit, "Satal wood" and Kawals' mushrooms (Elwin, 1964). Tribals of Kalahandi believe that their ancestors' survived by drinking the juice of the "salap" tree, after a catastrophe and therefore rivers are sacred. (Fernandes et al., 1988). Above all the Universal concept of "Vana Debta", the forest as a God, to be appeased, ensures the renewal of the species, while working a self imposed law against destruction of forests (ibid).

Tribal societies operate with an indigenous knowledge very effectively with the help of which not only they eke out a living but also solve the problems of their day to day social living. These healers occupy a significant socio-cultural and health related importance in their societies. They have their own knowledge in health care practices that are ecologically and culturally deep-rooted. The tribal health care system even now largely depends upon the locally available natural resources. Further the tribal healers are the potential resources of indigenous knowledge on health care, which is the only alternative to enrich the health care access of tribes.

The tribal people are well known for their "exotic" customs and practices. A substantial part of the tribal lore, myth, legends, folk tales, folk songs and folk dances reaffirms the concerns of a tribal community for its eco-system and culture.

It is not surprising to see that most of the forests are in tribal areas or vice versa. The reason is obvious. Generally, the tribal has a symbiotic relationship with the forest as they can not visualize a life without jungle that provides them food, fodder and fruits and have deep rooted sentiments and socio cultural needs (Jha and Sen, 1991). The subsistence characteristics of the way tribal people in the area make their living access to the natural resources is vital for maintaining and possibly improving, their livelihood.

FOREST POLICY AND ITS IMPACT ON TRIBAL ECONOMY

The drive towards better forest management and development led the National and State Governments to enact stringest laws and policies with regard to forest, there by gradually enhancing and monopolizing their power over it. Far from recognizing the relationship of tribals to the forests, these laws squarely blamed the tribals for its degradation. Moreover instead of legalizing existing rights and privileges of the tribals and the dwellers of the forests, the law deprived them of it. After the enforcement of forest protection Act, tribals who enjoyed vast rights and privileges over the forest has now limited use of it. Tribal Land rights are almost a non existent concept. The status of tribals with regard to land rights is clear from the fact that in government records, the tribal is an "encroacher" of the land he has been occupying through the ages. Whether he lives deep in the forest or in accessible hilly regions, he has no legal right over the place.

Collection of minor forest produce has become very difficult for tribals. Hard pressed to meet their needs for food from forests, the tribals have become more dependent on the local market (Hats). The changes have not only altered the tribals food habitats but also ethos of their life and culture. A study on Tribal Economy and its dependence upon minor forest produce and agro forestry system in six major tribal intensive districts of Orissa conducted by the center for Development Studies in its report, entitled "State of Orissa's Environment: A Citizen's Report"(CPSW..1994), reported that the seclusion of forest life is no more visible. The ecological interdependence of the tribals is declining. The tribals are being ousted perforce from their own homes. Now-a-days the very self sufficient economy of the tribal is undermined. Interferences of the intruders in the vicinity of the tribal area by the name of development have affected their local economy.

Incapable of bargaining for a better price for their own commodities has left the tribal economy virtually bankrupt. The phenomenon of developmental projects displacing people especially tribals, is clearly beyond doubt. But it is pitiable to note that in spite of providing so much of input for development and earning much for the state exchequer, the lot of the tribals has not improved. The bane of development projects has not only been in terms of displacement, but also in terms of the failure to ameliorate the lot of tribals. It is a paradox that the tribals are the poorest lot even though for centuries they have been occupying regions rich in resources.

The importance of the above mentioned discussion is that developmental planners, in order to widen their view had to be aware of not only epistemological aspects, but also of coso-molgies. In corporating the former in development planning implies incorporating cultures and culture perspective. The cultural categories that people use to classify and understand their environment should therefore be important to all those interested in its management. But environmental conservation is not a new concept for tribal people. By the name conservation the basic concern of the human being should not be forgotten. Eco tourism has been proposed by a number of interest groups as a new way forward for environmentally sustainable development. In particular, it asks, whether eco tourism constitutes genuine sustainable development with minimal negative social, political, environmental and economic impacts.

IMPACTS OF ECOTOURISM ON TRIBAL LIVELIHOOD

The impact of eco tourism on tribal life has caused much concern among anthropologists with a view to commercialization of their culture and economy. Several case studies show that the profit from such an enterprise goes to the ex chequer or the private parties and never to the local people. There is a strong lobby that feels that eco tourism has alienated the local people.

Negative Socio-economic and Environmental Impacts

- Tribals live very close to nature. Their 'economy' revolves around nature. They depend on forest for their survival. For promotion of eco tourism different codes of practice, Conservation Act will be implemented and interferences of Environmental Activists have deprived them from enjoying some of their customary rights over land and Forest resources which adversely affect their traditional economy.
- The relation between the tribals and the Forest is 'Symbiotic'. From the time immemorial the tribals occupy a central position in the ecological niche and the primary dependence of man on forest hints several strands of thought-ecological determinism, ecological possibilism and culture (man shapes his environment through culture). Their socio-cultural life is linked with forest. Many rituals are performed to worship the forest. By the name of conservation and economy the natural resources are being auctioned where the basic concerns of human beings are forgotten. This adversely affects their socio-economic condition.
- By the name of eco tourism the private investor will enter and occupy a key position in market economy. As a result, there will be a large-scale migration of non-tribals to the tribal areas, which would adversely affect their local economy.
- Increased tourism activities have both direct and indirect negative impacts on natural environment-air, water, noise pollution environmental degradation etc. Eco-tourism activities demand some structural construction-roads, hotel complexes; restaurants, which invade the natural beauty and there by affecting it.

- Tribals have their unique cultural identity. Regular interferences of intruders to the tribal life is unethical. Increasing flow of tourists to the tribal area has affected their eco-system and culture. Unfortunately the so-called eco-tourism has turned them into a specimen of zoo. Further the commercialization of different aspects of their culture in the name of eco-tourism both by the state and the local NGOs have created a condition of cultural genocide.
- Tribal lore, myth, legends, folktales and folk songs always have a unique position in their habitat and culture. The regular inflow of tourists into these natural habitats may destroy the fragile ecosystem and tribal culture.

Socio-Economic and Environmental Benefits of Ecotourism

- Tourism revenue generated can be channeled to the management of protected areas.
- It can diversify the local economy, particularly in rural areas where agriculture employment may be sporadic.
- It serves as an effective tool for visitors by creating awareness and helps in instilling in them a degree of concern for their ecosystem.

CONCLUSION AND RECOMMENDATIONS

Development is a concept of western society (Davidson, 2000; Escobar, 1995). What takes place now-a-days in the name of development can be viewed as "anti-development" because the process of development is always incomplete and relative. It is only growth-mania, belief in the commodifications and commercialization of economy and culture and vulgar display of consumer materialism. This kind of development has caused irreparable damage to the poor. It has caused the further marginalization of the already marginalized people and pushed to the periphery of the society. They are forced to pay the cost of the rich and the powerful people's urge and craze for so called development. The irony of the fact is that the cost is paid by the people who are least able to afford and who will not share the fruits of development.

In recent times the tribals have been exposed to various external forces and conditions that hamper their growth and development.

As more and more tribal dominated pockets are drawn to the process of modernization in the so-called development, there is the less and less social space for the tribals to participate in the process of decision making affecting their life. As a consequence, they have not able to protect themselves and preserve their rich cultural heritage. Both their cultural identity and human dignity have been under constant threat. The government policy of "integration not assimilation" is intended to live an autonomous cultural space for the tribals, while bringing them into political economy of the national mainstream. But such a policy has been problematic in its conceptualization and more so in its implementation.

Eco tourism or nature based tourism is already a popular concept, which is considered to be one of the growth areas of the tourism industry. But many people have different interpretations and connotations. It is like a big bag where people throw everything into it and create a big confusion in its understanding. It is time that we should have a correct and thorough understanding of the word eco tourism and bring much needed awareness.

To a host population eco tourism can be a mixed blessing. It is no doubt that tourism industry brings about changes in local communities. These changes are cultural, environmental and socio-economic. The introduction of tourism in the local communities affects the traditions and social life of villages. The tourists themselves become a social burden to villagers in the sense of introducing new kinds of behaviour, there by challenging traditional codes of behaviour.

Furthermore the eco tourism industry depends on the natural environment that the villagers traditionally depend on for their on-farm and off-farm activities. Introduction of tourism in the villages has brought about changes and natural resources have come to the access of villagers. These changes imply increased competition and degradation of natural resources. Ultimately this is threatening villager's livelihoods and the villagers ask for alternative means of income to meet their needs.

The eco tourism industry does create some jobs for the villagers. Hence, some have engaged them in tourism related business, either as their only source of income or as additional income generating activities. Therefore, job opportunities are beneficial for the livelihood of villagers. However, local employment effects from the tourism sector have been less significant than assumed. By the local population tourism is first of all appreciated as a new area for income generating activities in the informal sector. This new income activity which tourism implies, comes in addition to other more traditional activities and hence it is increasing the local risk diversification which is perceived very important in these communities.

The Draft National Environmental Policy 2004 is outlined to promote eco tourism through private investor ignoring the tribals who are most dependent on environment and natural resources. Environment is primarily for the people. It is the human element that gives meaning to ecology and the question of environment becomes relevant to society only to the extent that is viewed in relation to the people who are affected by it. Hence any development linked to environment has to be its relationship with tribal people who mostly depend on it for survival. The most unfortunate thing is that the tribal people who are the supposed beneficiaries of the eco-tourism are never consulted while designing or implanting such projects. It is an urgent need to work for the restoration of a balance between nature and human needs.

When government considers, eco tourism as a beneficiary industry, who actually benefits from eco tourism has remained an unresolved debate. Several case studies show that profit from such an enterprise either goes to state exchequer or the private investor and never to the local people. The question of sustainability arises when those who plan eco tourism activities without the involvement of the local people. Eco tourism can be sustainable if local people are taken into confidence. The local people should be made one of the important stakeholders in any eco tourism project. Most projects have failed to understand the concept that the local people have a better understanding of their eco system and can thus play a key role in conservation activities. More important than that they have an inborn sense of responsibility to protect the environment for the simple reason that their existence depends on it. So while introducing eco tourism in the region by the name of development, it is important to take into account the local, social, cultural and natural environments of tourism investments. Local livelihoods can only be secured by a higher degree of local participation in tourism planning. In the present context we must pay respect to Anthropologist Verrier Elwin's approach of isolationism "to allow the tribe man to live their life with utmost possible happiness and freedom and where missionaries of any sort were not to be allowed to break the tribal life". From the anthropological perspective any development for tribals to be implemented in the tribal areas should have a "tribal touch" or "tribal bias" which meant looking at things through tribal eyes and from the tribal point of view, respect and recognize their way of doing things. It can be concluded that for the name of eco tourism, tribal culture and tribal economy should not be exploited. If at all eco tourism is going to be promoted in tribal areas they should have the *right to decide their own priorities.* In case of requirement government, NGOs, Stake holders, funding agencies, private investors may act as facilitators.

Considering the afore-said background some suggestions can be proposed in the strategies of sustainable eco tourism development in tribal areas.

- Specific guidelines concerning eco tourism can be adopted using local talents, as no two places are alike. The module specially designed for eco tourism must be "site specific" and "selective" whether sanctuary, park, conservation projects etc.
- Local authorities and local communities can organize themselves to start the measurement and assessment before introducing the projects. The measurement progress can be related to social auditing and monitoring technique including indicators, targets, and sustainability reporting and eco foot-print analysis.

There should not be unrestricted entry into the tribal areas in the name of eco tourism. Agencies promoting eco tourism should take care to orient the tourist to interact with the environment and people without disturbing the local norms.

When the tribal areas are exposed to outsiders, certain items like tribal art and crafts as well as items not available else where are exposed to market forces. Situation needs to be created for the tribal people to sell their products directly without the intervention of the middleman, so that a proper regulatory mechanism can be obtained where, the local people can get the direct benefit from the tourist. Emphasize participatory approach, inviting participation of the people not only at the implementing level but also the designing of the same. The co-partners in the process of development should be equal partners. Local (tribal) educated youth may be appointed nature guides. Local people should be encouraged and expected to undertake leadership role in planning and development with the assistance of government, business, financial and other interests. Environmental Impact Assessment (EIA) study is a tool for any type of developmental project, so EIA must be conducted before introducing eco tourism. A steering committee may be constituted comprising the representatives from concerned Departments, District authorities, Stakeholders and community members The committee will be held responsible for constant monitoring of the process. Considering the environmental impact in the process, Environmental Management Plan (EMP) could be worked out for further course of action.

Ecotourism should try to avoid a clash between conservation and the people. The end benefit of ecotourism should go both to the community and the natural environment and it is imperative that local communities should be involved in the ecotourism activities. A networking and practical mechanism for tour operators should be adopted. They must provide visitors information and interpretation and services covering particularly, what to see, how to see, how to behave.

17

Tribal Problems and Remedial Measures

The whole of tribal phenomena is at the present moment going through a critical stage of transition. That the reaction of each tribe or group of tribes has not been same is bome out by the fact that there has been no uniformity in the population trends of various tribes. The tribal people are traditionally forest dwellers and they consider forest as their nourishing mother, which supply them food, fuel and fodder. But with the shrinking forest resources their hardship increases. Not only that they are coiled with various other maladies relating to impact of new revenue policies, the application of Indian civil and penal codes. Not only that, there has been contact with non-tribal people which has in most cases, due to the unscrupulous nature of the latter, led to indebtedness, land alienation and ever serfdom. Forced labour without payment of wages has been exorted by all kinds of overloads, including petty government officials. But unfortunately in the past, no concrete achievement could be made in solving the tribal problem inspite of the best intention of the government. The government machinery, the lack of enthusiasm to work among the tribes, the administrative bottle-necks at different stages of implementation of the scheme etc, where the major obstacles in getting the desired results in eradicating the tribal problem.

Before making any suggestion for removing the problems. I first discuss some of the major problems and the factors of the tribes in the study area, in briefly in this chapter.

The whole of tribal phenomena is at the present moment going through a critical stage of transition. That the reaction of each tribe or group of tribes has not been same is borne out by the fact that there has been no uniformity in the population trends of various tribes.

The tribal people are traditionally forest dwellers and they consider forest as their nourishing mother, which supply them food, fuel and fodder. But with the shrinking forest resources their hardship increases. Not only that they are coiled with various other maladies relating to impact of new revenue policies, the application of Indian civil and penal codes. Not only that, there has been contact with non-tribal people which has in most cases, due to the unscrupulous nature of the latter, led to indebtedness, land alienation and ever serfdom. Forced labour without payment of wages has been exorted by all kinds of overloads, including petty government officials. But unfortunately in the past, no concrete achievement could be made in solving the tribal problem inspite of the best intention of the government. The government machinery, the lack of enthusiasm to work among the tribes, the administrative bottle-necks at different stages of implementation of the scheme etc, where the major obstacles in getting the desired results in eradicating the tribal problem.

Main Factors of Tribal Problem

There are various factors responsible for the emergence of tribal problems. Chief among them as follows;

1. *Contact with outsiders*: The first cause of tribal problems is the contact of the tribal people with the outsiders. Due to this contact, tribal people have started to adopt the modern mode of living. They are adopting new ideas and practices consequently they are loosing faith on their own custom and traditions. This impact is responsible for their social and cultural destination.
2. *Exploitation by outsiders*: Moneylenders, traders and others often exploit the tribal communities, who come in the contact of outsiders. This exploitation has created the various social and economical problems.
3. *Policy*: The policy adopted by the Government towards tribal communities, is also one of the main cause of tribal problems. The administrative staff posted in the tribal areas, was totally unaquinted with the tribal customs and

traditions. Therefore sympathetic attitudes was not adopted towards the tribal people.

4. *Unapproachable habitation*: Generally the tribal people reside in those areas, which is unapproachable, or mountainous covered with dense forests. In those areas life is in a. primitive stage, means of transport and communication are under developed. Thus the primitive mode of life and geographical isolation of the tribal settlements have already created various problems.
5. *Christian missionaries*: The activities of the Christian Missionaries in the tribal areas have also created many problems.

Under the British administration some tribal areas were marked as excluded and partially excluded where no outsiders except the missionaries were allowed to go. The missionaries tried to convert tribal people into Christianity. Thus they tried to assimilate them into Christian society. This has created serious problem of cultural disintegration in the tribal society.

PROBLEMS OF THE TRIBAL COMMUNITIES IN THE STUDY AREA

There are various problems, which are confronted to the tribal communities. These problems have been summarized as follows;

Problem relating to poverty and backwardness

The poverty and indebtedness crisis comes to the forefront in tribal phenomena. It is though the tribal money lending system that the tribals are greatly exploited. The tribal economy is characterically subsistence type of economy and due to their type of economy and due to their somewhat traditional hedonistic type of worldview, they have meagre amount of saving. Naturally when they are in need of money, they are forces to depend on others particularly the non-tribal moneylenders. Under this system, a person who takes loan from moneylenders who are landowners is required to serve him like slaves according to the terms and condition of the debt. Quite often, if he is unable to repay the loan, it descends to his son or even to his third generation. Considering the degree of exploitation and seriousness of the problem, the 'Fifth schedule' empowers the Governors to take special action.

This system of forced bonded labors is in practice in many tribal settlements in the study area. It is a significant factor that the labor commission has taken steps to release the tribal children from the moneylenders or landowners in H.D.Kote taluk tribal settlements.

The lack of proper data in tribal indebtedness is a great hurdle. Although, some studies were made on tribal indebtedness long back about which some references are available in the reports of the commissioners for Scheduled Caste and Scheduled Tribes, Government of India, but no efforts have been made to get on up to date figures of tribal indebtedness at the micro-level or at the level of a particular ethnic group, so that immediate steps could be taken to remove the indebtedness.

In this study, it has been noted that the following factors are highly responsible for the tribal indebtedness:

- Merger income of the tribes.
- Ignorance of equitable price-system.
- Large-scale celebrations and festivals, especially on the occasions like birth, marriage, deaths etc.

Remedial steps

The fifth schedule of the constitution empowers the Government of a state to regulate the carrying on the business as moneylenders by persons who lend money to members of scheduled tribes in scheduled areas. In pursuance of this provision various State Government of have promulgated and enacted various laws and acts. Planning commissions Reports of the study team in the Tribal Development Programmes have summed up the situation with clarity. It is of the view that the attack on tribal indebtedness should be two-prolonged if it is to be effective and enduring. On the one hand long-term credit should be provided to enable the tribes to free himself from the strong hold of the usurious moneylenders and on the other hand improving his economic condition should raise his repaying capacity.

Co-operative credit should forthcoming in an adequate measure and the procedure has to be simplified to attract the tribes to take advantage of the credit facilities by the society. But the crux of the problem still remaining i.e. needs political will, commitment with honesty and integrity on the part of those who are supposed to be the protectors of tribal interests. In the absence of these, the coming regulations, provisions and schemes shall continue to be treated are mere pieces of paper.

Problem relating to Land Alienation

Land is the main stay of the tribes and more than 80 percent of them are dependent on agriculture and allied activities. Their economy is primarily agro-based. Land is the only tangible asset of a tribal family. Tribes have emotional attachment to land. However, with the opening up of the tribal areas, the tribal land is being alienated to the non-tribals. During 2000-2001, the cultivatable land was 71.04 percent in the study area. Remaining 28.46 percent of land has been alienated from non-tribais, followness and land has not been suitable for cultivation. Consequently a large number of tribal cultivators have been rendered landless labourers. The agricultural labourers were 30.49 percent in 1999-2000. Which is more than other occupationist. It shows the decrease in the number of cultivators (13.15 percent) and increase in the landless labourers is also an indication of the distributing trend of land alienation. The incidence of land alienation varies from region to region. Apart from the field survey (1999-2000), about 43.36 percent of the tribal households have been affected by the land alienation.

Ariofiler tendency fiignfrgfited by the survey was that in certain areas, although the tribes are the official owners of the land, the non-tribes have become its virtual owners. This problem is very acute in the Melukamanahalli and Moguvinhally tribal colonies of Gundlepet taluk. The study of land alienation among the tribes has not been studied in detail. A case study of land alienation in rural areas was conducted by tribal welfare organizations which brought to light the nature and agencies associated with land alienation. Regulations under which cases of alienation of last few years could be opened. Mere legislation is not enough, it is essential, firstly, to study the various patterns of land alienation covering the entire Scheduled Areas.

Remedial steps

Taking into consideration of the three bases of land alienation of tribal people viz.,

(a) The lacunae in the laws,

(b) The ignorance of the tribal people, and

(c) The complicated legal procedures to be. Many state governments have been doing something to help the tribal. Before looking into various legislative steps taken to safeguard tribal interest in land, it will be in the fitness of things to study the crucial recommendations of Scheduled Areas and Scheduled Commissions, 1961, on the basis of which various states passed a number of legislations.

Some of the important legislation enumerated may be divided into two types; one is protective intended to prevent non-tribal acquiring tribal land and the other deals with question of land reforms. Both type of legislation has fallen short of the needs of the situation. The protective legislation could not be of much use as no sufficient arrangement for alternative credit was made. The land reform legislation could not help the tribes to a great extent because it was too intricated for him to fallow. But the most important factor in this state of affair is the lack of will and integrity on the part of adrninistration to ameliorate the sufferings of the tribal masses. Hence the vested interests exploiting the tribes have been and still are in a position to carry in their nefarious and criminal activities some time with the help of indifference of officials and some time with the active connivance of officials at lower levels. Due to their growing contacts with radical political elements and sincere pro-peasantry cards of radical organizations, an explosive is fast growing up.

Problem of shifting cultivation

There are some minor tribes who even today practice the shifting cultivation about which some reference have already been made in chapter III of this study.

Under shifting cultivation a vast area of forestland is selected for shifting cultivation and the trees and plants of the forest are cut and burnt. Under this system a vast track of forest land is destroyed by cutting the forest, which is not only harmful for them but, it also creates ecological imbalances in the region. Therefore, the government has banned the practice of shifting cultivation, and hence, those tribes who have been practicing the shifting cultivation, are facing seasons problems.

Remedial steps

The report of the Scheduled Area and Scheduled Tribes Commission, 1961, has recommended some important suggestions for implementation. Some of the major suggestions are as follows:

1. The total replacement of this system, if it come at all, will be a long-term process. In the mean time there must be a proper course to regulate it, experiment with it, improve it and try other workable alternatives of shifting cultivation tied up with the tribal way of life and it is much more than a technical problem pertaining only to improve the techniques of forming and agriculture.
2. Shifting cultivation may be converted into terraced cultivation wherever the hill slopes are ideal and the soil is sufficiently deep. This holds out immerse possibilities of improvement, and in many places throughout the tribal areas it has been taken up with enthusiasm.
3. Establishment of agncultural/calonies may also be an effective solution to wean away the tribes from shifting cultivation to sedentary cultivation. Under this programme, tribal families may receive some cultivable wasteland, financial assistance for the construction of houses, reclamation of the land, bullocks, agricultural implements and high yielding verity seeds etc., Facilities for communications, drinking water, irrigation, medical aids, etc., were also proposed. However the following measures should be provided for sedentary agriculture and to improve the agriculture in tribal areas.
   - Setting up of a demonstration farms
   - Distribution of improved agricultural implements, manure, fertilizers and HYV seeds
   - Extending credit facilities.

Improved agricultural implements distributed to tribes have been such as would meet their approval.

Problem relating to forests

The tribal people everywhere have lived in intimate relationship with forest and it has been linked with their very existence due to obvious factors. Forests have been sufficiently sustaining their inhabitants. Forests are their abode and source of livelihood. They get food, fruits, roots and honey from the forests. Thus the traditional tribal economy is mainly built around the forests. Also many flora and fauna have been the objects of worship for the tribes due to one reason or the other. And not less important is the fact that the tribal people achieved a harmony between their lives and nature through the forests.

But the forest policies are not encouraging the tribes in many aspects, like growing, collection of forest produces formation of settlement in dense forests, land reforms etc., Although the forest policies are not to disturb for tribal livelihood, it vary from region to region. The forest Act has enhanced harassment and exploitation of the tribal people. The forest offices have to work in an extremely different physical condition, therefore they are given legal protection of a far reaching character. Under this guard the poor tribes have certainly been placed at the mercy of the forest officials. The unscrupulous officials among them have been playing havoc with their lives and harass them at the slightest pretext.

Collection of minor forest products an is not at all likely to hinder the forest either in its growth or preservation. But at certain places the tribal inhabitants are not allowed to avail of this concession. Instead the forest officials get it collected and uses it for their benefits.

The large-scale deforestation leads to ecological imbalance, nonavailability of forest products and non-availability of raw material for forest based industries. It also influences the rainfall, atmospheric quality and causes floods, landslides, soil erosion and other natural disasters. The ultimate victims are the tribes.

Remedial Steps

There are several developmental policies launched by the government for the upliftment of tribes. But the basic defects of these policies are that they are fonnulated without taking into consideration of the ground realities existing in the different remote tribal pockets. A particular policy may be suitable in one area and may not be suitable to other areas, as there are differences in ecological situations. Therefore, the utmost care has to be taken while formulating to preserve the policies. The forest and creation of new forest areas to compensate the loss of green cover in the last few decades. Tribes all along used to get most of their necessities from the forests. They never know the technique and care of raising different plant species. They have been exploiting the forest resources since time immemorial. Therefore, necessary training should be important to them in this respect by forest extension workers at the village level. In tribal areas, the villages are usually very small and consists of only 10 to 15 houses. In such cases a group of villages should be checked together and the centrally located village should be considered as the key or model village. The village level workers both government and voluntary organisations can effectively make use of these model village for the implementation viable forest policies in the field of forest, healthcare, education etc., for the comprehensive development of the tribal areas.

Lack of Educational Facilities among the Tribes

In the past, education was one of the main considerations for hierarchical classification in the Hindu society. Lack of it is largely responsible for the exploitation and pitiable plight of the tribes. Education is the key catalyst for the development of human resources. For the tribes, education is the pivot on which success depends. Education disseminates knowledge. Knowledge gives inner strength, which is very much essential for the tribes for attaining freedom from exploitation. Due to ignorance arising out of illiteracy, the tribes have not been able to take advantage of new economic opportunities. Opening of the tribal areas in the work of developmental process have brought in juxtaposition two distinct value systems one based on tradition and ignorance and the other on technologies and innovations. Harmonious syncronisation of the two systems is essential for the development of tribal society. In this process, education will play a key role. The most important aspect of education in the tribal areas is that of informing the community about the new innovations is science and technology as well as the development in economic and political fields. Thus education must be fruitful to the tiibal people for the enliancement social and economic status.

Recognising the importance of education, the constitution has made specific provisions in Article 15(iv) and 46 for promoting education among the scheduled tribes.

According to 1981 census the literacy among the scheduled tribes was 14.58 percent against the general literacy of 31.34 percent. A comparative position of literacy percentage among the general population and scheduled tribe population as per 1981 and 1991 census is given in table; Although, literacy among the tribes has increased considerable over a period of one decade it is still far below the general level. The position of female literacy is really a matter of concern. The rate of tribal literacy varies from region to region. There are some important causes for slow-progress in literacy among the tribes. Such causes are to be maintained specially in this study and they are follows.

Social Factors

More allocation of funds and opening of new schools are not far behind in providing education to the tribal people. A few obvious factors play the dominant role in this respect. Formal education is not very much necessary for the members of tribal societies to discharge their social obligations. Economic Factors

Some Economic factors, too, are responsible for the lack of interest shown by the tribal people in getting education. Since most of the tribal people are living in utter poverty under subsistence economy, it is not easy for most of them to send their children to school thus losing two healthy hands in their struggle for survival. Verrier Elwin (1963)' very appropriately sums up the situation in the following way: "For a tribal family, to send its grown up girls or boys to school, is essentially a matter of economics; and entails dislocation in the traditional pattern of division of labour many parents cannot just afford to send their children to school."

According to the present system of education any economic benefit that a tribal child can bring to his family will be only after 10 or 15 years of schooling. The parents have neither the patience nor the foresight to wait for such a long period. They can however, be convinced easily if the education for them can be made productive right from the very beginning.

Lack of Interest in Formal Education

Sharma (1976) has rightly pointed out that the "urban middle-class-orientated education system has got superimposed on the entire nation both in terms of its structure and content." In many states tribal children are taught the same books, which from the curriculum of non-tribal children of the urban and rural areas of the rest of the state. Obviously the content of such books rearly appeals the tribal children who come from different cultural backgrounds. Stories of scientific and technological progress, founders of modern India, history and geography of the country etc, of course, from necessary part of any curriculum but the situation demands that their education should start with teaching of demography, history and ecology of their own region. National consciousness should not be imposed from above or outside, but they should be made aware of it in a systematic manner.

Dedicated Teachers

Lack of dedicated teachers is one of the major reasons for the slow growth of education in tribal areas. Most of the teachers employed for imparting education to the tribal children show little appreciation of the tribal way of life and value system. They approach the tribal people with a sense of superiority and treat them as 'savage' and 'uncivilized' and hence fail to establish proper rapport with their students.

Lack of Facilities

Nature of habitat of tribes is also responsible for the slow growth of education. Most of the tribal villages are scattered. This entails long travels to attend schools. School builds in some cases, also plays an important role in the growth of education among the tribal folk. Due to mismanagement, buildings and some times financial constrains are seldom suitable to run an educational institution. Another factor related with the problem is the number of teachers. Most primary schools in tribal areas are "single-teacher managed whose presence in the school is more an exception than a duty." Over burdening may be one of the possible factors for this state of affairs.

Remedial Steps

Thus, it is evident that the tribes of India and Mysore district are in different stages of development and naturally the problems of education related to those living in different stages of development, will be different. Therefore in order to eradicate illiteracy among the tribes of India and more particularly in the study area an integrated approach at the micro level will highly useful.

Problem of Unemployment and Migration

The tribes have migrated to different Tea gardens in search of getting employment since many years. Tribal agriculture and allied activities are seasonal and it depended on rainfall. Jemikuruba and Soliga tribal communities in Gundlupet and Kollegal taluks were migrated to Kerala and nearly by regions. It supports the idea that in search of employment, the tribes have been migrating to far of places and whose descendants hardly return to their homeland.

The tribal labourers are hard working and they migrate with their family members. Except the kids, both husband and wife will be engaged as labourers. But they are very much exploited by the contractors who pick them from their villages and take them with proper transportation arrangements to the work spot either at tea garden or at the factory site or at the road sites where constructions are going on. But the most unfortunate part of these migrants and the newly recruited tribal labourers is that they become unemployed as soon as the earthwork is over or the construction is completed..

In this way, the tribes of Chamarajanagar district, particularly the members of the minor tribes like the Jenukuruba, Kadukuruba etc., who are illiterate and innocent exploited more by the contractors. This is not, however, true with the members of other minor tribes because, they live in the inaccessible jungle areas and they depend totally on the forest. They are therefore, untapped by the labour contractors.

Remedial Measures

Planning to solve unemployment should be formulated tat micro or local level with due weightage for specific problems of a particular place because of variation from one tribal area to other. Priority should be given to local tribes while appointing forest workers, watcher, guards etc., by the concerned authority. But unfortunately this is not being done due to administrative bottle-necks.

Housing Problem

The problem of housing for tribal people has to be viewed in the following aspects;

- Shelter
- Sanitation
- Aesthetic tastes, and
- Comparative costs

Environment occupies an important place in the above-mentioned aspects of housing scheme of tribes. Variations in climate coupled with differences in economic condition of tribes and structure posses a number of problems. All the tribal communities are fortunate enough that they have been endowed with natural bounty. They are living in eco-friendly surroundings. Melukamanahalli, Kaniyanapura, Baragi, punajnur, Budipadaga etc., are the eco-friendly regions, which are situated at Bandipur National Park in Gundlepet taluk and Biligirirangaswamy temple wild life sanctuary in yalandur taluk. Many of the tribal families have developed a real pride in having good houses. On the other hand many of tribal families completely lack in house sense.

Government has built and granted the houses for many needy tribes. But these houses are unimaginative and misfit for the life style of tribes this is happening because some one will prepare model of the tribal house who generally never exposed to the tribal surrounding and ascertain the facts that what type of houses are convenient to the tribes.

Hence. Any scheme of housing without considering tribal consciousness is bound to meet the disapproval of the perspective in habitants of the houses.

Moreover the forest department is also coming in the way of solving tribal housing problem. The unimaginative and reckless bureaucracy has posed many restrictions on use of forest for house construction by the tribes. The tribes can collect the house construction materials from the forest by obtaining permit. But it is very difficult for a tribal to obtain permit from forest officials for obvious reason, which are known to all. Hence the entire forest policy has become counter productive as for as the tribal welfare schemes is concern.

Remedial Steps

Housing schemes for many impoverished tribal groups like soiigas, Junukurubas, Kadukuruba, Hakkipikki, Meda etc, in the study area are to be implemented on top priority. In this direction many of the housing schemes like Ashraya schemes, Ambedkar vasathi scheme etc launched by the state government are to be effectively speeded up for the benefit of tribes. But it should be kept in mind that even an explosive house is useless for tribal people if they feel uncomfortable in unfamiliar house surroundings.

HYGIENE AND HEALTH PROBLEM

The tribes are suffering from many chronic diseases and the most prevalent which take heavy tolls of them are water-borne. This is mainly because of poor drinking water supply. Even though plenty of water is available in and around the tribal settlements, generally it is dirty and contaminated. Diarrhea, dysentery, cholera, etc, are often the results of this situation.

Apart from this deficiency of certain minerals and other elements are also in the water the causes for such diseases. In B,R Hills tribal region, sickle cell Anemia and Tuberculosis are common among many tribes. Added to this maternity problem is very serious in tribal colonies. Even today maternity work is done by the tribal women. One tribal woman known as Jallesiddamma (staying in Yarakanagadde podu) doing wonderful service in this field and she has been awarded by the government for her health service. Her service is incorporated with all the tribal settlement in B.R.Hills. Hence she has been recognised as "Soliga Thayandira Herige Doctor" (Doctor of maternity) by the tribes and local people.

Remedial Steps

There has been considerable expansion of medical and public health facilities for tribes during the last two Five Year Plans. Health centers are being opened with qualified doctors and with good medical facilities inspite of this it is very difficult to trace the improvement in their health condition.

Due to the lack of transportation and communication in tribal areas, they feel reluctant to go to the hospitals that are located far away from the tribal settlements. Hence primary health center should be established in the remote tribal areas with expert doctors.

Moreover, mobile medical service has to be introduced for serving the tribal community. This will play a long inning in solving the health problems of tribes.

Apart from this non governmental organisations should be effectively involved in solving health problems of the tribes in this regard, Jayavijayam tribal hospital located in B.R. Hills in association with VGKK, a non-governmental organization, It serves the tribes with modern medical aids at free of cost. It also bridges the tribes with communicate the modern medical field.

Problem of Communication

Tribes are living in isolation for centuries. The main reason for this isolation is lack of communication. The problem of communication in tribal area may be examined from two angels: (1) needs of the tribes and (2) developments of the tribal areas as a whole. It has been ascertained that the communication facilities may not always prove to be a blessing to the tribal society. It also facilitates the entry of perspective exploiters and other anti-social elements from non-tribal areas, who takes the advantage of innocence of the tribes and exploit them in a variety of ways. But at the same time the march of events cannot be halted and the needs of defence, industrialisation, mining and development can also not be ignored.

Many tribes are good cultivators of some growing variety of crops. With a good network of roads and communication, their production will be able to reach far off markets and they can fetch handsome returns. In the absence of adequate means of communications and transportation. They have to sell their produce at comparatively low prices and are easily brow beaten by the local non-tribal traders to sell their produce at throwaway prices. Mukti colony is far away from the marketing place (24 Kins) in Gundlepet taluk without proper transportation and communication. Hence, they are forced to sell their agricultural produce at low prices after harvesting and it is inhavitable for them due to lack of storage facilities.

Post and Telegraph services are slowly percolating into the remotest tribal areas, but unfortunately the mail carriers and staff are not regular employees of the post and telegraph department. Beduguli is one of the best examples for this inadequate service.

Remedial Steps

Development of communication no doubt will provide good opportunities for the tribal folk to develop national consciousness and a healthier intellectual horizon. Also the importance of communication in the realm of tribal economy hardly needs an emphasis but what has to be guarded against the danger of greedy and cunning, elements from the plains invading tribal areas to exploit the poor tribes. The best solution in such a situation seems to be taking all sorts of precautions for safeguarding the tribal interest before throwing any of their areas open.

Rehabilitation problem

As an example of general failure in tackling the problem of rehabilitation of displaced tribals more than 2000 tribal families in Karemala, Melukamanahalli, Moguvinahalli, Mukti colony, Maddur colony and Mukalli colony in Gundlupet taluk, Muneswara colony and Srinivasapura colony in Chamarajanagar taluk, Anehola, Kavalichalla dam in Kollegal Taluks are distributed. These are all rehabilitated tribal colonies. Government has taken care to fulfill their basic needs, but, it is far away from expectation and situation is not happier for the reason mentioned below.

Very few of them have taken advantage of agricultural facilities offered in their habilitation camps. The attraction of easy money and industrial employment are too strong, especially for young tribes.

The alternative land offered by the authorities for cultivation is not irrigable and cannot offer adequate return to maintain the families on it.

The cash compensation given to tribes seldom-utilised far productive purposes. It is almost invariably used for daily living expenses until alternative employment is found.

Remedial steps

In the sense of rehabilitation Government should provide only fertile land for cultivation in and around accessible regions. If once the tribes are rehabilitated, the government should provide them clean drinking water, electricity, communication and as much as possible the irrigation facilities for agriculture under the Ganga Kalyana Scheme, Nooru Bavi (Hundred Well) Scheme etc.,

PRESENT PROBLEMS OF THE TRIBES IN THE STUDY AREA

Tribals are generally addicted to alcohol i.e.Impact. They have traditionally been brewing liquor from ragimalt. Arrack entered into the tribal area in the eighties after they came into closer contract with the outside world. Arrack being sold by private persons in the tribal areas, who takes the contract from the government by the auction. This provides an easy access to tribal men and they have become regular drinkers and addicted. Many times tribal women protested. Against liquor sale in the tribal areas. It is evident from Punajanur tribal settlements. The tribal women protested and sat 'dhama' along with tribal development agencies and women co-operative societies in B.R.Hills and Punajanur state forest dwellers on 8th August 2001.

'Kaniyan' a community known for their expertise in astrology belongs to the Schedule Tribes in Kollegal taluk and their cousins in neighboring Yalandur belong to the other backward caste and are called 'Kaniyar'. There is no proper reason why this is so. Meanwhile people in Kollegal Kaniyan tribal community have been provided houses under Ashraya scheme and have been given other facilities by the state government. The Yalandur taluk counterparts envy their cousins and curse their fate.

Despite these disparities Kaniyans were migrated from Kerala and settled in different parts of Tamil Nadu, parts of Hassan district, Malavalli taluks of Mandya district, T. Narasipura taluk of Mysore district and Kollegal and Yalandur taluk of study area. They recall that after the reunification of states in 1956, the people in Kollegal taluk were called as Kaniyan and included in the list of Scheduled Tribes. Those settled in the neighboring Kannada speaking taluks and other parts were included in the list of other backward communities through they were called Kaniya or Kaniyas by their fellowmen.

A tribal settlement called Upkar colony is 65 Kms away from Gundtepet taluk. It is a rehabilitated settlement of about 10 tribal families from Bandipur National Park, When they are evicted from the forest, government has provided 150 acres of barren land to them which is not more suitable for cultivation. Now, this settlement is suffering from many problems and the most serious are unemployment, and the attack by wild animals. Hence, the tribes have already inclined to migrate to nearer village called Kundakere (40 kms away from the tribal settlement) mostly in search of employment.

Tribes are making their effort to grow crops from their smallholding. But hardly there are getting there because of wild animals minace. The elephants are destroying most of the crops before harvesting. Maddur, Mukti colony, Moguvinahalli colony, Budipadaga, Beduguli etc., are the worst affected areas. Loss of life is also common phenomenon by wild animals attack two soliga men were recently killed by the elephant in Srinivasapura tribal colony near Punajanur in the study area.

The entire Jurisdiction of the study area is the hunting ground of forest poacher/smuggler Veerappan. Veerappan activities affect the tribes both for and against.

To put and end to his unlawful activities a special task force has been formed by the government and this STF is operating in the jungle from quite a long time. In the name of operation Veerappan the tribals settlements are being disturbed tribal men and women are being torched mentally and physically. This situation is unbearable in modern civilized society. Hence, for all said and done to uplift the tribal communities the solved the early end to this problem and it will be a success if all of us stretch our helping in this regard. Conclusion

Tribal problems are very complex in nature and studies are not free from this. These problems are intensified due to imposition of tribal organisation act. This act establishes the legitimate demand of tribes on their environmental problems. With the same time the socio-economic backwardness forced them to depend on available resources around their settlements. The situation clearly indicated that the under development of tribes in the society. The government has implemented various measures, but these measures are not sufficient and effectively reaching them both in time and place to over come the problems.

At the same time tribal sub-plan has taken a lot of developmental programs at various fields like agricultural, economical, social forestry etc., by which new economic prospects will be created in this area. But, until and unless the tribal areas are provided with suitable infrastructure. These induced measures would not be a successful. The main hindrance of attaining the useful infrastructure is the gradual exploitation by means of indebtedness and land alienation's.

To compact this major problem some socio-political organisations by non-tribal people of this area are working among the tribal people. They are mainly concentrating to create the awareness about their problems. They also high light the problems to the local authorities and provide some sorts of help to the badly needy tribal people. These groups sometimes conduct exhibition in tribal areas for understanding the tribal problem in a simple way to attract the tribal people. However, this problem can be treated as prime constraint for their upliftment in historical process and if this problem is ignored what will happen in future is just a question today.

18

A Comparative Review

INTRODUCTION

The Convention No. 107 & Convention No. 169

The ILO Convention on Indigenous and Tribal Populations of 1957 (Convention No. 107), and its accompanying Recommendation No. 104, were adopted in 1957. Described as a 'seminal instrument', the convention is regarded as having "remained unique in international law as the only comprehensive international statement of the rights of indigenous and tribal populations and of the duties of states toward them from 1957 until 1989". Convention No. 107 covers a wide range of issues, including land, recruitment and conditions of employment, vocational training, handicrafts and rural industries, social security and health, administration, education and means of communication. Reflecting the 'mainstream' ideas of the time, the Convention sought to provide basic safeguard measures for indigenous and tribal population groups.

In 1989, ILO Convention No. 107 was revised and the more progressive ILO Convention on Indigenous and Tribal Peoples of 1989 (Convention No. 169) was adopted. The latter was adopted in order to incorporate current perspectives on the protection of the rights of indigenous peoples. It dismissed the 'integrationist approach' of Convention 107, recognising instead the right of indigenous and tribal people to exist as a distinct "peoples", should they wish to do so. This paradigm shift was the result of several developments, especially since the 1980s. These include the increasing awareness and mobilization among indigenous peoples and NGOs, the conclusion of a study on discrimination against indigenous populations by United Nations Special Rapporteur Jose R. Martinez Cobo in 1986, increasing international acknowledgment of the need to end such discrimination and facilitate the equal application of universal human rights standards to indigenous peoples, and the beginning of a drafting process of a global declaration on indigenous peoples' rights, in 1984. Since 1989, further ratifications of Convention No. 107 are no longer open, but it continues to be valid for those countries that have ratified it, and have chosen not to ratify the later convention of 1989. By 1989, nineteen countries had ratified ILO Convention No. 107, while twenty countries have ratified Convention No. 169, including only two countries in the Asia-Pacific region, Fiji and Nepal.

Background to Ratification of Convention No. 107

The ILO Convention No. 107 was ratified by Bangladesh in 1972, a year after it gained independence from Pakistan. However, as a part of former Pakistan, the convention had applied to Bangladesh even prior to independence. Pakistan – a signatory to the Convention-had ratified the instrument in 1960. However, apart from the periodical observations of the ILO's Committee of Experts – upon which body is vested the responsibility of monitoring the implementation of the ILO conventions-and a number of ILO mission visits to Bangladesh, direct dialogues and contact between the ILO and the Government and indigenous people of Bangladesh have been quite limited, including at the country level.

ILO Convention No. 107 Article 1.

1. This convention applies to:

(a) members of tribal or semi-tribal populations in independent countries whose social and economic conditions are at a less advanced stage than the stage reached by the other sections of the national community, and whose status is regulated wholly or partially by their own customs or traditions or by special laws or regulations;

(b) members of tribal or semi-tribal populations in independent countries which are regarded as indigenous on account of their descent from the populations which inhabited the country, or a geographical region to which the country belongs, at the time of conquest or colonisation and which, irrespective of their legal status, live more in conformity with the social, economic and cultural institutions of that time than with the institutions of the nation to which they belong.

2. For the purposes of this Convention, the term *semi-tribal* includes groups and persons who, although they are in the process of losing their tribal characteristics, are not yet integrated into the national community.

Two, among other, important features of the aforesaid criteria may be noted. First, that although the Convention refers to three types of population groups, and provides different criteria to help identify them, when it comes to the application of its provisions to the aforesaid three groups, the Convention does not distinguish between them. In other words, *the provisions of the Convention apply equally to members of all three groups – indigenous, tribal and semi-tribal-without any distinction*. Second, while the aforesaid criteria have, on the whole, proved to be useful in identifying indigenous and tribal population groups, they carry a number of underlying assumptions and implications that are disrespectful towards the people concerned, and are also out of step with the progressive development of international human rights law, and the realities of contemporary indigenous society. In his report on *Discrimination against Indigenous People,* UN Special Rapporteur Martinez Cobo provides several criteria to identify indigenous peoples, and these have gained increasing respect among indigenous peoples and human rights scholars. Cobo highlights continuity with pre-invasion and pre-colonial societies, the concerned groups' membership of non-dominant sectors of society, their ethnic identity, close ties to their ancestral territories and the continuation of their social institutions and legal systems, among other things. A formal definition of the term has never been provided, including in the *UN Declaration on the Rights of Indigenous Peoples,* since such definitions are feared to lead to the exclusion of legitimate groups, rather than actually help identify them. The *UN Declaration on the Rights of Persons Belonging to National or Ethnic, Religious and Linguistic Minorities,* for example, also does not include a formal definition of minorities. Although indigenous peoples and minorities groups share some common experiences, the former do not wish to be considered as minorities because "the historical, traditional and cultural circumstances of indigenous peoples are different to that of minorities".

ILO Convention No. 169 Article 1 1. This Convention applies to:

(a) tribal peoples in independent countries whose social, cultural and economic conditions distinguish them from other sections of the national community, and whose status is regulated wholly or partially by their own customs or traditions or by special laws or regulations;

(b) peoples in independent countries who are regarded as indigenous on account of their descent from the populations which inhabited the country, or a geographical region to which the country belongs, at the time of conquest or colonisation or the establishment of present state boundaries and who, irrespective of their legal status, retain some or all of their own social, economic, cultural and political institutions.

2. Self-identification as indigenous or tribal shall be regarded as a fundamental criterion for determining the groups to which the provisions of this Convention apply.

3. The use of the term peoples in this Convention shall not be construed as having any implications as regards the rights which may attach to the term under international law.

Populations and Peoples under Conventions No. 107 & 169

In a general sense, there are, of course, extremely important differences, legal, political, and otherwise, between the meaning of the terms 'populations' and 'peoples'. For the purposes of the two above-named ILO conventions (107 and 169), however, these distinctions are of relative unimportance on account of the qualified use of the term 'peoples' in the 1989 Convention. Article 1(3) of the Convention (No. 169) reads; "The use of the term "peoples" in this Convention shall not be construed as having any implications as regards the rights which may attach to the term under international law". This qualification was not acceptable to many indigenous people that were involved in the drafting process, but the qualified reference remains in the Convention. The *Declaration on the Rights of Indigenous Peoples* uses the term peoples without any qualification. Therefore, the use of the term 'peoples' is in conformity with current progressive thinking on the subject.

The Integrationist Emphasis of Convention No. 107 and Current Discourses on Integration

Among the reasons behind the revision of Convention No. 107 were its integrationist approach and patronizing attitudes. The convention "presumes that [indigenous and tribal peoples] will disappear as separate groups once they have the opportunity to participate fully in the national society, and it attempts to ease the transitional period".However, in the later and more progressive ILO Convention No 169, the reference to such phrases as "less advanced stage" has been dropped, as was the reference to "semi-tribal" groups. Moreover, the criterion of self-identification, another important tool for identification, has been added. It is no longer regarded that the loss of "tribal characteristics" by the peoples and population groups concerned is either inevitable, or desirable, as was assumed by the drafters of the Convention in 1957.

In fact, more and more population groups the world over are increasingly asserting their identity as indigenous peoples, having suffered through, and survived, some of the worst forms of discrimination and colonization known to humankind. Again, in line with current development thinking of that time, which assumed that decisions regarding development were the concern of the State rather than of the people most affected, the drafters of Convention No. 107 are not known to have either consulted indigenous peoples about the provisions of the Convention, or sought their informed consent prior to its adoption.

In contrast, the drafting of Convention No. 169 did involve a number of indigenous leaders and experts, although it too is regarded as falling below the minimum standards for indigenous peoples, for example, in comparison to the *Declaration on the Rights of Indigenous Peoples*. However, equally importantly, the contents of Convention No. 169 also reflect the paradigm shift in the international developmental and human rights contexts, which have increasingly acknowledged that the views of indigenous peoples regarding their rights and welfare not only needed to be accounted for in the development process, but that these peoples have the right to be a substantive part of all relevant decision-making processes.

Such progressive views have been manifested in the work of several international human rights mechanisms, including the UN treaty-monitoring bodies, the Human Rights Committee and the Committee on Economic, Social and Cultural Rights, the Office of the UN High Commissioner for Human Rights, the former UN Commission on Human Rights and its then subsidiary body, the UN Working Group on Indigenous Populations. Similar views are also reflected in the policies of international lending or financial institutions (multilateral development banks) and bilateral development agencies, including the World Bank and Danida (one of the development agencies of the Government of Denmark). Lastly, but not the least, some very progressive and landmark shifts in legal thinking can be seen in such studies as that of UN Special Rapporteur Martinez Cobo mentioned above, and most importantly, in this case, in the deliberations at International Labour Conferences under the auspices of the ILO.

A handbook on the ILO and Minorities and Indigenous Peoples, published in 2002 by the Minority Rights Group International, states the following: "At the time of its adoption the approach towards indigenous peoples was paternalistic, with integration being a major aim...... The ILO no longer takes this approach, and integration is not an issue of concern or of interest to the supervisory bodies." On account of the above directional changes, including the views of the ILO's supervisory bodies, it is incumbent upon all who are concerned with the implementation of Convention No. 107 to re-interpret many of its provisions in tune with current and progressive discourses, both on moral, and practical, grounds.

THE INDIGENOUS PEOPLES OF BANGLADESH

'Indigenous', 'Hillmen', 'Aboriginal', 'Tribal', 'Adivasi' under Bangladeshi Law & Practice

The issue of the identity of the indigenous peoples of Bangladesh has led to much debate and controversy, and on occasions has brought indigenous leaders and government officials into sharp disagreement. The aboriginal groups in the Chittagong Hill Tracts (CHT) are generally known as 'Pahari' (meaning hillpeople), or as Jumma (from the common tradition of swidden or 'jum' cultivation). The plains aboriginals, particularly those in the north-western greater Rajshahi region, used to be generally known as *Adivasi,* meaning aboriginal or indigenous. This term had little currency in other parts of Bangladesh until the 1990s. However, ever since 1992, when the *International Year of the Indigenous Peoples* was declared by the United Nations, more and more indigenous peoples, both from the CHT and the plains, have started to increasingly refer to themselves as *indigenous* in English, and as *Adivasi* in the national language, Bengali.

Government perspectives on the issue, however, are varied. The Ministry of Foreign Affairs, for example, has at times preferred the terms 'tribe' and 'tribal' ("upajati in Bengali) to 'indigenous' and 'adivasi'. Non-indigenous functionaries at the district and sub-district levels too were generally seen to prefer the terms "tribe and "tribal;" or their Bengali equivalent, 'upajati', but the term Adivasi is being used more widely, especially in the plains districts.The legal situation, on the other hand, is quite pluralistic, and reflects, in its totality, the currency of all the terms preferred by government officials and indigenous peoples combined. Recent legislation seems to favour the terms "tribe/tribal". A 1995 law – primarily a finance law, but also containing specific references to the exemption of income tax payments by indigenous people in the CHT – uses the term "indigenous hillmen". Other laws that were framed in earlier periods, such as the CHT Regulation of 1900 (Regulation I of 1900), use the terms "indigenous tribe" and "indigenous hillmen" interchangeably. The term "indigenous hillmen" is also echoed in several references to the Finance and Income Tax Acts in the schedule to the 1900 Regulation, and in several correspondences of the National Board of Revenue. Furthermore, in a case in the High Court Division of the Supreme Court of Bangladesh in 2000, the court took cognisance of the fact that the petitioner was an "indigenous hillman" of the CHT.

In the plains regions, the East Bengal State Acquisition & Tenancy Act, 1950 uses the phrases "aboriginal castes or tribes" to refer to several groups who now identify themselves as indigenous or Adivasi. In the National Poverty Reduction Strategy Paper adopted by the Government of Bangladesh in 2005 ("PRSP-I"), the term "Adivasi/ethnic minorities" was used. In recent years, this was the first time that Bengali equivalent of indigenous or aboriginal appeared in a formal government document. In the PRSP-II, adopted in October 2008, the terms "indigenous communities" and "indigenous people" were both used.

In several governmental radio advertisements on non-discriminatory treatment for all ethnic and religious groups in the follow-up to the 2008 parliamentary elections, the word "adivasi" was used, alongside the Bengali equivalent of minorities ("shongkha loghu"). In a proposed amendment to the Union Councils Ordinance, 2008, a provision was sought to be introduced to maintain reserved seats for *Adivasis* in union councils ("union "parishads") in areas outside the CHT where there was a significant Adivasi population. If the amendment had been put into effect, this would have been the first time in recent years that the Bengali equivalent of the terms indigenous and aboriginal was used in a statute.

Regards the terminology used, the private press and media too display divergent attitudes, but on the whole, most tend nowadays to be respectful towards the views of the peoples concerned, and 'indigenous' and 'adivasi' are generally preferred over 'tribes' or 'tribal'. Similarly, more and more prominent figures of Bangladeshi civil society are being seen to use the terms indigenous or Adivasi. For example, a political party leader and now Member of Parliament and government minister, recently wrote; "The diversity of our culture due to the presence of indigenous communities is providing extra vigour to the national fabric of Bangladesh. Moreover, indigenous people are the original inhabitants of our country. So, they have the same right we have over Bangladesh, if not more."

Throughout this study, wherever the context directly relates so, the author will use the terms 'indigenous', 'tribal', 'peoples', 'populations' etc. as they occur in Conventions No. 107 or 169, or in the concerned international or national instrument under reference. However, in general, he will use the term 'indigenous' rather than 'tribe' or 'tribal' for a number of reasons, including those referred to above, to mean all three groups: indigenous, tribal and semi-tribal. Firstly, this is more in conformity with the wishes of the peoples concerned as expressed over the years in formal declarations and statements. Secondly, the terms 'tribe' and 'tribal' have come to acquire certain pejorative connotations in Bangladesh and elsewhere over the years, being associated with "primitive", "backward" and so forth. Thirdly, there is a firm legal and political basis for using the term 'indigenous', along with the similar term 'aboriginal', since both occur in formal and subsisting Bangladeshi laws. This is also consistent with current discourses in the progressive development of human rights law, including the *UN Declaration on the Rights of Indigenous Peoples*. However, as far as the application of the provisions of this convention is concerned, the issue of formal definition, including how the government chooses to identify or define any or all of the three groups concerned (indigenous, tribal and semi-tribal), is irrelevant. Instead, the aim of Convention No. 107, as is also the case with Convention No. 169, is to address the specific vulnerability and socio-economic situation of the population groups and peoples concerned.

The Indigenous Peoples

Estimates on the number of indigenous peoples in Bangladesh, and their population, tend to vary a great deal. Without accounting for differences in the way the names of some peoples are spelt, the number of the different peoples has been estimated as twelve, twenty, twenty-eight, thirty, forty-five and forty-six. The reasons for the different numbers includes the multiple names by which the same group is known by different peoples, the different ways of spelling the names of the groups, the categorization of a sub-group as a separate group itself, and so on.

The differences are particularly sharp for the peoples of the plains, with the number clearly rising in recent years, indicating a conscious decision of some peoples to increasingly identify themselves as indigenous. In recent years, members of tea estate workers – largely descendants of migrant workers of indigenous descent from neighbouring areas of India, who remain among the most socially, economically and politically marginalized sections of Bangladeshi citizens – have also chosen to participate in forums of indigenous peoples.

One writer sees this rising tendency towards self-identification as indigenous or Adivasi as a result, among others, of "international ethnic politics", and "efforts of some Hinduized groups to disassociate themselves from the caste system", particularly in the plains districts. While this may be partly true at best, acute discrimination against plains Adivasis is surely also a major factor behind the denial of Adivasi identity up to the 1990s. The problem in Bangladesh is not so much one of pretenders seeking indigenous status, but more one of denial of identity and rights of genuine indigenous people. Sometimes inter-faith conflicts within the same ethnic group coupled with discriminatory attitudes of district and sub-district government officials has even led to the denial of certificates of Adivasi origin – required for admission into Adivasi/ tribal quota seats of certain educational institutions – on account of religious affiliation.

indigenous peoples, and this has also been reflected in recent legislation. Some groups with very small populations that live mostly around the three district headquarters, who are not part of this list, including the Gurkha, Ahomia (Assamese) and Santal, for example, have also demanded the inclusion of their names as indigenous or tribal, a demand so far denied. Table 1 in the Annexe shows the different peoples according to different classifications along with estimates of their population as available in the official census of 1991. The total indigenous ('tribal') population of Bangladesh was estimated at 1,205,978, out of a total population of 111,455,185, making up 1.08% of the national population. The tribal population of the CHT was 514,805 and the tribal population of the plains was 691,173, respectively, 42.67% and 57.33% of the total tribal population of the country.

Indigenous People, In-migration & Minoritization

In some parts of the country, such as the CHT, Bengali-speaking people did not permanently reside at all, until British colonization in the 19 century. As in the case of neighbouring Northeast India, soon after the annexation of the CHT into the British Indian Empire in 1860, in-migration of members of peoples not regarded as "native" to region was strictly controlled through special laws and regulations, including the Inner Line Regulation, 1873.These laws were later repealed in the case of the CHT (while they still remain valid in parts of Northeast India) and the population of Bengali-speaking people began to rise in the region, gradually at first, and then rapidly after the independence of Bangladesh in 1971. In particular, migration reached very high rates in the early 1980s, on account of a population transfer programme that rehabilitated between 250,000 to 450,000 landless Bengali people (predominantly Muslim), in the CHT through direct government sponsorship. The last official census with ethnic break-up was the 1991 census. This estimates the 'tribal' population of the CHT to be about 51% of the total population of the region, although there is good reason to believe that this is an underestimate. In contrast, the indigenous population made up 98% of the population of the region in 1872 (1872 official census).

With regard to plains Bangladesh, the nature and extent of minoritization is generally even sharper. The clearest example is greater Mymensingh district in north-central Bangladesh, a large part of which was, until 1962, included within a "partially excluded area", denoting an area predominantly inhabited by indigenous peoples, including the *Mandi or* 'Garo', *Hajong, Koch* and *Hadi.* Here too, the non-indigenous population has grown rapidly since the 1950s through overt and covert encouragement of, and patronage by, government leaders and officials. Today, most indigenous communities in the plains form small pockets or enclaves in an otherwise Bengali-populated area. This means, among other things, that these peoples' electoral strength does not usually allow them to elect an indigenous representative to local government bodies, let alone to parliament, with some very limited exceptions.This obviously impinges upon their opportunities of 'participation in elective institutions' as mentioned in the Convention.

As in the case of the CHT, the 1991 census figures for the plains indigenous peoples are also believed to be 'obvious' underestimates, particularly in the northwestern Rajshahi division. The aforesaid demographic changes have had far-reaching, and largely negative, consequences on their political, social and economic integrity, and their cultural identity. The situations in Mymensingh and the CHT in particular are clearly of the sort that the Convention sought to avoid, when it spoke against the 'artificial assimilation' of the population groups concerned. Population transfer in both places was contrary to the non-discrimination clauses of the Constitution of Bangladesh, the Convention on the Elimination of Racial Discrimination (CERD), ratified by Bangladesh, and other general principles of international law concerning population transfer.

THE LAWS OF THE PLAINS AND INDIGENOUS PEOPLES

The only substantive law that deals with the plains indigenous peoples is the East Bengal State Acquisition and Tenancy Act, 1950. Another law that mentions plains indigenous peoples – along with those in the CHT – is the Drugs and Alcoholic Substances Control Act, 1990. In addition other laws that affect plains indigenous peoples, although they do not directly refer to these peoples, include the Vested Property Act, 1974 and the Forest Act, 1927, including subsidiary rules framed under the last-named Act. In addition, there are the personal laws of the different peoples, which are regulated largely by traditional customs and practices. Therefore, the overall situation of extremely limited legislative, administrative and supervisory measures concerning indigenous peoples' issues of the plains region illustrates a state of affairs that falls far below the standard that was contemplated in Part VII (Article 27) of Convention No. 107. This contrasts with the situation in the semi-autonomous CHT region in the southeast, where one sees much more elaborate legislative, administrative and supervisory mechanisms concerning indigenous peoples' issues.

The East Bengal State Acquisition and Tenancy Act, 1950

The East Bengal State Acquisition and Tenancy Act, 1950 (Act XXVIII of 1950) is the major law regulating land administration in the plains. A small part of this statute – at Section 97(1)(2)-deals with restrictions on the transfer of lands of "aboriginal castes and tribes" to any other than "[aboriginals] domiciled or permanently residing in Bangladesh".

Although somewhat paternalistic, the aforesaid provision of the law seeks to prevent land alienation among indigenous peoples. This law predates the ILO Conventions No. 107 and 169, and ILO Recommendation No. 104 and concerns situations that are precisely of the nature that are contemplated by article 13 (2) of Convention 107, which provides that "arrangements shall be made to prevent persons who are not members of the populations concerned from taking advantage of these customs or of lack of understanding of the laws on the part of the members of these populations to secure the ownership or use of the lands belonging to such members." Similar provisions are contained in Article 17(3) of Convention 169, and Recommendations 5(1) and 6 of ILO Recommendation 104, which are compatible with the more elaborate provisions of the Declaration on the Rights of Indigenous Peoples (especially articles 25-30).

The application of the law, however, is unequal across the plains. Except in parts of greater Mymensingh district that were formerly 'partially excluded areas', there are widespread reports of non-implementation of this law. In Mymensingh, there is an understanding between the district administration and the *Tribal Welfare Association* – a social organization that goes back to the British period – whereby the district authorities do not allow land transfers by Mandi (Garo) and other aboriginal peoples without first consulting the association. Similar but somewhat less institutionalized practices of consultation are evident in

Dinajpur district, but not in other parts of Rajshahi division. Such practices are consistent with the provisions on collaboration and participation of indigenous peoples, as contained in article 5 of Convention No. 107. More elaborate guidelines on legislative and executive measures in cooperation with indigenous peoples are contained in Part VIII (Article 33) of Convention No. 169.

The situation of land alienation in the Rakhaing-inhabited south-central Patuakhali-Barguna region is also a matter of particular concern. During British times, and perhaps even on to the early period of Pakistani rule, special officers – generally syled as *Welfare Officers* – used to bear special responsibility to protect the welfare of indigenous peoples, including in Mymensingh and in the Patuakhali-Barguna region, but there is no such office in the aforesaid places any more. It is reported that in Dinajpur district, the post of Welfare Officer still exists, and yet seems to have little or nothing to do with the welfare of the local indigenous people.

Despite its rather paternalistic setting, and its limitations in dealing with fraudulent and violent land-grabbing not involving land title transfers, the Tenancy law of 1950 nevertheless has the potential to minimize, if not totally stop, land alienation of the plains indigenous peoples. Certain provisions of the Act could perhaps be more proactively promoted and invoked in favour of indigenous peoples. Reforms to include provisions on repossession of dispossessed lands, and/or the formation of a body to provide expeditious remedies for land alienation cases would perhaps improve the situation a great deal. It cannot be said whether such reforms will eventuate in the near future. It is therefore incumbent to make the best possible use of this law. As it were, for reasons that are unlikely to be related to the indigenous peoples, the 1950 law also happens to be one of the laws – and in fact, the first-named law-that is specially protected by the Constitution of Bangladesh in its First Schedule. This means that the constitutionality of the law – as in the case of the other laws included in the Schedule – cannot be questioned in court on grounds of being contrary to the fundamental rights clauses of the constitution. This is an important safeguard, considering that various comparable laws of special application in the CHT have undergone legal challenges in the superior courts.

The Drugs and Alcoholic Substances Control Act, 1990

The Drugs and Alcoholic Substances Control Act, 1990 (Act XX of 1990) exempts "tribals" – both in the plains and in the CHT – along with tea estate labourers and the castes of "Muchi" (cobbler), 'Methor' (sweeper) and 'Dom' (fisher-people or cremation ground cleaners) from criminal prosecution for consuming traditionally brewed or distilled alcoholic beverages. Although seemingly a mundane matter, this is an important recognition of the cultural rights of the indigenous peoples, whose religious and cultural practices and social customs often included, and still include, the use of traditionally brewed liquor. This provision may be seen to cover situations contemplated by Articles 7 (2) and 8 of Convention No. 107. Article 7(2) provides for the retention of customs (and institutions) of indigenous peoples, while article 8 provides that government authorities and courts dealing with penal matters involving indigenous peoples should bear in mind the customs of the people concerned.

The Vested Property Act, 1974

In the plains, another tool used by influential and unscrupulous land-grabbers has been the Vested and Nonresident Property

(Administration) Act, 1974 (Act XLVI of 1974), generally known as the *Vested Property Act*. This Act is based upon a number of laws framed in 1965 and thereafter, to deal with properties of people who went over to India, with which country Pakistan (Bangladesh was then the eastern wing of Pakistan) was at war with in 1965. The 1974 Act has been indiscriminately used against different minority groups, including indigenous peoples in the northwestern Rajshahi division, and against Mandi (Garo) people in north-central Bangladesh,among others. Although the law has since been repealed, certain loopholes stand in the way of redressing wrongs perpetuated in its name. The return of lands of indigenous people taken over by invoking the Vested Property Act is one of the 9-point demands of the Jatiyo Adivasi Parishad, the main organization of the indigenous peoples of the northwest. The discriminatory application of the aforesaid Act is clearly contrary to Article 13 (2) of Convention 107, which provides for arrangements to "prevent persons who are not members of the populations concerned from taking advantage of ..... lack of understanding of the laws on the part of the members of these populations". This law was amended in 2009 by the Caretaker Government of Bangladesh through the passage of an ordinance. The amendment sought to simplify the process of de-listing of lands unfairly or baselessly included within the list of vested property.

The Forest Act of 1927

In addition, there are a number of other laws, which do not specifically mention the plains indigenous people, but are of direct relevance to their land and resource rights. One such law is the Forest Act of 1927 (Act XVI of 1927), which regulates the manner of the administration of forest areas and forest produce. Various provisions of this Act, particularly section 28, concern customary land rights of indigenous people Section 28 (1) of this Act empowers the government to "assign to any village community the right of Government to or over any land which has been constituted into a reserved forest". This provision is compatible with article 11 of Convention No. 107, which provides for the recognition of the collective and individual rights of indigenous and tribal people over lands traditionally occupied by them. Since a significant number of plains indigenous people live within reserved forest areas, particularly in the northeastern Sylhet division, in the north-central greater Mymensingh area, and within Chittagong and Cox's Bazar districts of Chittagong division, this provision has important implications for their rights.

In some parts of Sylhet division, village communities of the Khasi people have secured short-term written agreements over use of reserved forest land, but on the expiry of the lease period, the same were not renewed.Other communities of indigenous peoples in greater Mymensingh, greater Chittagong and Cox's Bazar continue to live within reserved forest areas, but no written agreements are known to have been entered into with them, unlike in Sylhet division. Their existence and livelihoods in these areas are therefore even more precarious. In November 2009, the Ministry of Environment and Forests formed a committee to consider the framing of rules on Forest Villages in accordance with section 28(1) of the Forest Act. The same committee was also entrusted the

responsibility of drafting a Traditional Forest Dwellers (Recognition of Rights) Act. The framing of the proposed rules is therefore crucial to both protect the rights of the forest dwellers but also to bring forth cooperation between the Department of Forests and forest-dwelling communities. The concerned rule-making prerogative is vested upon the Ministry of Environment & Forests, and does not require formal legislation by parliament. It is not known, however, whether the ministry is actively considering reforms in this sector.

The Social Forestry Rules

Another law that affects the land rights of indigenous peoples is a delegated law, namely, the *Social Forestry Rules of* 2004, passed in accordance with the aforesaid Forest Act of 1927 [sections 28A(4) and 28A(5)]. These rules contain detailed provisions for social forestry projects, and "ethnic minorities" (a phrase that would include most groups legally classified as indigenous, tribal or aboriginal) are among those that are to be given priority in selection as beneficiaries of the project (along with landless people and "destitute women"). The aforesaid provisions are compatible with article 15 of Convention No.107, which concerns prevention of discrimination against indigenous and tribal workers, and special measures for their recruitment. They are also compatible with the provisions of ILO Convention 111, which concerns the elimination of discrimination in employment.

Customary Personal Laws

Bangladesh, as in Pakistan, India and Malaysia, has no uniform civil code of general application to govern the family laws of all its citizens. Instead, rules concerning marriage, divorce, separation, maintenance, advancement, child custody and inheritance, and other related matters, are regulated in accordance with the ethnic or religious affiliations of the citizens concerned. Thus, while Muslim Law governs the family laws of Muslim Bangladeshis, and Hindu law governs the family law of the Hindu Bangladeshis, the rules of family matters of the indigenous peoples are governed by their customary laws, which are largely unwritten, generally irrespective of their religious affiliation or spiritual beliefs. This is also the case with the hill peoples of the CHT. Personal laws of most Bangladeshi peoples – whether indigenous or Muslim or Hindu-tend to be generally discriminatory against women. Apart from the peoples among whom Christianity has taken firm roots, polygyny is practiced to a certain extent, and women are generally excluded from inheritance rights, either wholly or partially.

The noteworthy exceptions are in the case of the Mandi (Garo) and the Khasi, both of whom follow matrilineal traditions, where generally, only the women inherit landed property. The express or implied recognition of the indigenous peoples' customary laws is compatible with articles 7(1) and 7(2) of Convention 107, which deals with customary laws, customs and institutions of indigenous and tribal people. However, the presence of practices that are discriminatory against women is contrary to article 7(3) of the Convention. This article seeks to prevent situations in which the application of customary law may infringe upon rights that are otherwise available to other indigenous people as citizens of a state. Therefore, since the Bangladeshi constitution prohibits discrimination against women (articles 27 and 28), these constitutional provisions may be invoked to protect indigenous women's right to equality and to the equal protection of the law; the same would be an instance that was contemplated by the provisions of article 7(3) of Convention 107. The provisions of the

Convention on the Elimination of Discrimination against Women (CEDAW), which has been ratified by Bangladesh, is als0o a very relevant instrument in this regard. It is therefore vital for the protection of the rights of indigenous women in Bangladesh to consider the ambit of article 7(3) of Convention 107 in line with the progressive thinking on women's rights as acknowledged by the provisions of the Constitution of Bangladesh and those of CEDAW.

Personal Laws of Adivasis and Justice Administration in the Plains

As mentioned in chapter 3.6 above, the personal laws of indigenous peoples of Bangladesh, both in the plains and in the CHT, are regulated by their own customs, traditions and practices. In addition, the CHT also has its own justice system, including traditional indigenous chiefs (rajas) and headmen, whose justice administration and other functions are formally recognised by law. In contrast, the self-governing systems and representative institutions of the plains Adivasis have long ceased to be recognized by the Government, although their functions nevertheless continue, and most personal law disputes are still resolved at the level of traditional and other community leaders rather than in the government courts of justice. In 2008, a proposal was tabled before the Advisory Council of the Caretaker Government of Bangladesh to amend the Union Parishad Ordinance of 2008 to reserve seats in union councils for Adivasis of the plains. Unfortunately, the amendment did not eventuate. Had it taken place, the amendment could have been a major catalyst in strengthening the role of the plains Adivasis in local self-governance, reversing a trend of centuries, whereby their self-government institutions and justice systems came to be de-recognised and marginalized.

Otherwise, although their traditional elders and –leaders still exercise significant authority in the resolution of disputes among their communities, on account of the absence of any formal recognition by law, their authority has been severely eroded over the years. Reforms to the Union Parishad laws and to the Gram Adalat Court Act of 1976 (Village Courts Act) could provide a formal role to Adivasi community leaders in the administration of justice on personal law matters. The latter law provides for arbitration headed by the Union Council chairman, including nominees of the litigants, whose decisions have the force of law.

THE LAWS OF THE CHITTAGONG HILL TRACTS

In contrast to the plains, there is a large body of law in the CHT that relates to the unique administrative system in the CHT, some of which recognise the primacy of the indigenous peoples of the region. These laws can be broadly divided into two groups: those of the period before 1997-the year of the signing of the CHT "peace" Accord – and the laws of the post-1997 period. The first-named group primarily includes the CHT Regulation, 1900 and laws specified in its schedule (which apply to the CHT to the extent and in the manner described therein), and the Hill District Council Acts of 1989. The second groups includes the three practically identical laws of 1998, which amend the Hill District Council laws of 1989, the 1998 law that introduces the CHT Regional Council, the 2001 law on the *CHT Land Dispute Resolution Commission*, and lastly, the law on reforms to the civil and criminal justices system, passed in 2003. In addition, there are the customary and largely unwritten family laws of the different peoples, which are administered locally by the chiefs and headmen.

The CHT Regulation, 1900

The CHT Regulation of 1900 (Act I of 1900), is undoubtedly one of the most important laws of the CHT. It "lays down a detailed policy for the general, judicial, land, and revenue administration of the region and defines the powers, functions and responsibilities of various officials and institutions...... [It] stipulates the manner and extent of the application of other laws to the region, many of which apply only to the extent that they are not inconsistent with the Regulation." Thus it functions in the nature of a constitutional legal instrument, which vets the application of other laws to the region, although it no longer enjoys any formal constitutional status. The Regulation implicitly accounts for the practice of customary law, and therefore, lays down guidelines on the manner of regulation of land and resource-use patterns, rather than explicitly mentioning those practices and related rights. It has been said that "[the Regulation] was not intended to be a declaratory instrument that sought to identify, define and declare various customary rights and privileges, but a regulatory law that sought to regulate already existing rights…". The Regulation thus provides a simple system of administration of justice and recognizes the authority of three traditional chiefs and about 380 mauza headmen.

In 1989, when the unrest in the CHT was still continuing, and before the "Peace Accord" of 1997 was signed, the Government of Bangladesh introduced a legal package to introduce district-level councils in the CHT, and to repeal the CHT Regulation. However, the law intended to repeal the Regulation was never put into effect. Instead, the CHT Accord of 1997 and the CHT Regional Council Act, 1998 provide that the CHT Regional Council will advise the Government of Bangladesh to remove inconsistencies between the CHT Regulation and the Hill District Council

Acts, 1989.

The 1900 Regulation draws upon several special regulations used by the British colonial authorities in other parts of their then empire. Foremost among these are the Chin Hills Regulation, 1896 (Regulation No. V of 1896), formerly applying in some of the Chin-inhabited areas of Burma ('Myanmar') and to some parts of Northeast India, and the Daman-i-Koh Rules, hitherto applying to parts of Santal-inhabited territory in India, including what now forms Jharkhand state. As with other comparable special regulations of the colonial period, on the one hand, the 1900 Regulation affirms and acknowledges certain rights of the hill peoples. On the other hand, it also limits those rights by retaining supreme authority upon the government. In the case of Rangamati Food Products v. Commissioner of Customs & Others in the High Court Division of the Supreme Court of Bangladesh, while discussing the question of the applicability of tax laws to the region, the court declared that the CHT Regulation, 1900 was a "dead law". However, there were other judgments of the High Court in which the Regulation was regarded as a fully valid law, and therefore, a special bench was constituted to hear three related cases on the CHT and provide a ruling on the validity of the Regulation. The CHT Regional Council, the Chiefs and the headmen presented memoranda to the then deputy minister of the Ministry of CHT Affairs, urging steps to defend the Regulation. The deputy minister (also the MP from Rangamati) in turn, wrote to the Attorney General advising him that the Government of Bangladesh should take a stand in favour of retaining the CHT Regulation, 1900, citing the importance of the role of the chiefs, headmen and karbaries (village elders), the need to protect the rights of the different peoples of the CHT, related constitutional provisions, the likely negative impact on the legislative prerogatives of the CHT Regional Council, and the government's desire to soon put into effect the 2003 amendment of the CHT Regulation to reform the system of administration of justice in the CHT. The CHT Regional Council, the three Chiefs and headmen filed separate applications to be added as parties in the matter. Soon afterwards, the Government of Bangladesh filed an appeal against that portion of the judgment of the Rangamati Foods case that declared the CHT Regulation, 1900 to be a "dead law", presumably acting on the basis of advice received from the CHT minister. Consequently, the three cases before the Special bench were adjourned. Moreover, the effect of the Rangamati Foods judgment with regard to the validity of the CHT Regulation was stayed by the Appellate Division of the Supreme Court pending the outcome of the *Leave to Appeal* petition.

Several provisions of the CHT Regulation are compatible with Convention 107, or with those of the supplementary Recommendation 104, while some go beyond Convention 107. Some of the latter are closer to the provisions of Convention 169, and some are even comparable to the provisions of the UN Declaration on the Rights of Indigenous Peoples, which is regarded as an instrument that generally recognizes a higher standard of rights than that of ILO Conventions No. 107 and 169. For example, the provisions of the 1900 Regulation with regard to the offices of the chiefs and headmen (Rules 39, 40, 47, 48) demonstrate a situation of the nature that was contemplated by article 7(2) of Convention 107, which seeks to allow indigenous people to "retain their own customs and institutions". Similarly, several provisions in the Regulation concerning land-including rules 34, 41, 41A, 42, 43, 45B and 50 (1) – expressly or implicitly recognize the individual and collective rights of the CHT indigenous people over swidden, grazing and forest lands traditionally occupied by them, as contemplated in articles 11 and 13 (1) of Convention 107. Likewise, the provisions of rule 34(14) of the Regulation that restrict the inheritance of CHT land by non-residents are comparable to the provisions of Article 13(2) of Convention No. 107, which seeks to prevent land alienation of indigenous people at the hands of non-indigenous people through unfair means. Other relevant comparable provisions include rule 40 of the Regulation, which empowers traditional chiefs and headmen to try disputes among their people by invoking customary law. These are consistent with the provisions of Article 8(a) of Convention No. 107, which provides that "the methods of social control practised by the populations concerned shall be used as far as possible for dealing with crimes or offences committed by members of these populations". Then there is rule 38 of the Regulation that recognizes the duty of the chiefs to look after the education and health needs of their people, which is compatible with articles 20 and 21 of Convention 107, although the latter provisions more substantively address the responsibility of the government, rather than that of indigenous leaders or officials.

The Hill District Councils Acts of 1989

The Hill District Council Acts of 1989 (Acts XIX, XX & XX of 1989) established three identically-empowered councils at the district level, reserving two-thirds of their seats and the office of the chairperson, for 'tribals'. According to this Act, the number of seats for the members is to be allocated according to ethnicity and, ostensibly, the relative population of the different ethnic groups. The councils were provided limited authority over a number of district-level government departments. These laws were amended in 1998 in accordance with the provisions of the CHT Accord of 1997, whereby the councils were provided more authority and autonomy than before, through an increase in the number of transferred subjects, and enhancement of their authority over land administration, local police, budgets, and other matters. Each council is to have 33 members, excluding the chairperson, and the chiefs have the prerogative of attending any meetings of the concerned council, although they cannot vote. The councils are to be ultimately elected by tribal and non-tribal permanent residents of the concerned districts. However, apart from the first and rather controversial election in 1989, the councils have until now been administered by government appointees. At the time of writing, the councils have one chairperson and four members each, all men, who were appointed by the interim Caretaker Government of Bangladesh. Dissatisfaction has been expressed on several occasions over the non-representation and under-representation of women, in the interim district councils, even during the earlier periods under BNP and Awami League. Similarly, complaints have been made about under-representation of different indigenous groups as well. It is to be seen whether the newly-elected Awami League-led government provides justice in this matter.

The CHT Regional Council Act of 1998

The CHT Regional Council Act, 1998 (Act XXII of 1998) is another law that resulted from the 1997 Accord. It established a regional council for the CHT, to be indirectly elected by the members of the three district councils. As in the case of the district councils, the office of the chairperson and that of two-thirds of the members of the regional council are reserved for 'tribals'. It has 24 members excluding the chairperson. It is to be ultimately elected by an electoral college consisting of the 99 district councillors and the three chairpersons of the hill district councils.However, no elections for the regional council, or for that matter, for the district councils, have been held (except for the district councils in 1989). Pending elections, the members of the CHT Regional Council are all government appointees.

The CHT Regional Council exercises supervisory and coordinating authority over such matters as development activities, general administration, the hill district councils, local government councils, the CHT Development Board, and customary law and social justice, thus accounting for a wide array of administrative and developmental functions. One of the most crucial features of the Act is the recognition it provides to the region as a "tribal-inhabited area", and the prerogative of the council to advise the Government of Bangladesh on legislation concerning the CHT. However, in practice, the Regional Council has been unable to effectively play its expected role, as aspired for during the signing of the Accord, for a variety of reasons, including non-devolution of the requisite authority upon it, as claimed by the chairperson of the council, and the over-bureaucratized and departmentalized structure of the council.

Personal Laws & Traditional Justice Systems

As in the case of the plains, the customary personal laws of the different peoples in the CHT also vary from people to people. There are, however, two important distinctions, among others, between the plains and the 'hills'. Firstly, matrilineal traditions are less common in the CHT than in the plains. One section of the Marma people, particularly those living in the southern CHT, stand out as the only indigenous group in the CHT that allows its women to inherit some landed property, as of right, although not in an equal manner with its men. The second major difference is that while in the plains the self-government systems of the indigenous peoples, including their juridical systems, have been long deprived of formal state recognition, the customary justice system administered by chiefs and headmen in the CHT is still recognized by law. The customary law practices and the continuing functions of the traditional chiefs and headmen in the administration of justice, and in other spheres of administration, are compatible with the provisions of articles 7 and 8 of Convention No. 107. Article 7 acknowledges the need to give due regard to customary laws of indigenous peoples in defining these peoples' rights and duties, and likewise recognizes the right of the peoples concerned to retain their own customs and institutions. Article 8 gives importance to adherence to customary law in the administration of criminal justice and penal provisions.

Major infrastructural, economic and other developments, including the expansion of the road network and communications systems, the spread of education, growing urbanization and other related developments have triggered important occupational and other social changes in the CHT over the years. Polygyny, for example,is clearly on the decline, especially among the Christianized groups. There are strong demands among indigenous human rights and women's rights activists to completely outlaw polygyny, and to remove all other practices discriminating against women, including on inheritance and child custody. Other major challenges being faced by the CHT indigenous peoples with regard to their customary personal laws – which substantively follow oral traditions-is the question of whether to go for partial or full codification, and how to face the difficulties in the interface between traditional courts and state courts, and that between customary law and statute law. The Supreme Court of Bangladesh has strongly acknowledged the importance of protecting the customary laws of the CHT.

One other challenge to customary law is the situation of indigenous migrants from the CHT who now live and work in the metropolitan centres of Dhaka and Chittagong. There are an estimated twenty-five to thirty thousand workers in factories in these two cities, at least half of whom are women, who have to deal with their customary law matters (marriage rites, family law disputes, etc) without having access to traditional authorities (chief, headman, karbari). In a number of meetings called by the leaders of voluntary civic associations of these migrants, it was proposed that the city equivalent of village elders or 'karbaries' be nominated by the workers and formally appointed by the traditional *rajas* (Circle Chiefs). An understanding on this matter has been reached between these communities and the Chakma Chief, who represented the three chiefs in discussions with the indigenous migrant leaders.

OTHER POST-ACCORD LAWS

The CHT Land Commission Act of 2001

Apart from the regional and district council laws, the *CHT Land Commission Act*, 2001 (Act LIII of 2001) is the most important post-1997 law. This law introduces a body to provide expeditious remedies to the longstanding problem of land-related disputes, especially those between hill people and Bengali-speaking population transferees of the 1980s. Although called a 'commission', its functions will be more in the nature of an adjudicating body's, rather than that of a commission of enquiry. In order to ensure a fair process, the Act ensures that seven of its nine members are indigenous leaders. The commission is also obliged to take into account the "laws, customs and usages" of the CHT. Although the council has been set up, it is not fully functional yet, since the chairperson of the CHT Regional Council, and all or most of the other indigenous members, do not agree to start work until the discrepancies between the Act and the provisions of the CHT Accord are removed through legal amendments. Among these are the virtual veto powers provided to the commission's chairperson, in the event of disagreements between the commission's other members and uncertainties regarding the jurisdiction of the commission over forest land and seasonally cultivable plough lands of the Karnafuli reservoir area known as "fringelands".Moreover, the office of chairperson of the commission is also vacant, due to the death during office of the last incumbent, Mr. Justice Mahmudur Rahman. The newly-elected Government is actively considering the appointment of a new chairperson of the commission.

The CHT Regional Council has put forward its proposed recommendations to successive governments, including to the interim Caretaker Government of Bangladesh in 2008. In November 2008, the Ministry of CHT Affairs forwarded the recommendations of the CHT Regional Council to the Ministry of Land, so that the latter could place the matter before the Advisory Council of the Caretaker Government for legislation. For reasons that are beyond the scope of this paper to discuss, the amendment did not come through, despite a clear understanding on the issue between the CHT Regional Council and the Ministry of CHT Affairs.

After the present Awami League-led *Grand Alliance* government came into power, a number of developments suggest that the desired amendments might be brought about soon. These include public statements of the present Minister-of-State for the CHT Affairs Ministry, Dipankar Talukdar, MP and the Law Minister, Barrister Shafique Ahmed, among others.

In the event that amendments proposed by the CHT Regional Council are accepted by the government and passed into law, the following positive changes would be effected: (i) the work methods of the commission would become more inclusive and participatory (by removing the virtual veto powers of the commission's chairperson and by providing him a casting vote only in the event of a tie), (ii) the uncertainties over the jurisdiction of the commission would be clarified (by redefining "fringelands", for example), (iii) the workload of the commission would be lightened through delegation of primary investigation and enquiry (except final decision-making) to members or officials of the commission, and so forth. However, even if the commission does start its work in the next year or so, it is feared that it may yet take several years before all the concerned disputes before the commission are resolved, on account of evidential difficulties, time constraints, the unavailability of class actions under Bangladeshi law, and other such factors. Nevertheless, if the commission were to start its work soon, this would remove one of the most important obstacles towards peace-consolidation and justice in the CHT. It may arguably also be a unique example of restitution and justice in land disputes involving indigenous peoples worthy of being replicated elsewhere in the world. This would be in conformity with the spirit of the UN Declaration on the Rights of Indigenous

Peoples, which provides elaborate provisions on recognition of traditional land rights, restitution of alienated lands and a fair and inclusive manner of providing redress for land-related disputes, among others.

The CHT Regulation (Amendment) Act of 2003

The CHT Regulation (Amendment) Act of 2003 (Act XXXVIII of 2003) deals with the transfer of jurisdiction over the administration of civil and criminal justice – formerly vested upon civil servants at the district and divisional levels – to judicial officers under the *Ministry of Law, Justice and Parliamentary Affairs*. This does not affect the existing functions of the traditional chiefs and headmen in dispensing justice on tribal customary laws, which have been expressly excluded from the jurisdiction of the new courts ("except the cases arising out of the family laws and other customary laws of the tribes of the districts of Rangamati, Khagrachari and Bandarban respectively which shall be triable by the Mauza Headmen and Circle Chiefs").

Furthermore, these courts are obliged to try matters in accordance with "laws, customs and usages of the districts concerned", a clear and formal acknowledgement of customary law and practice, apart from personal laws of the indigenous peoples.

This law, which was passed in 2003, was finally put into effect in August, 2008, during the tenure of the caretaker Government of Bangladesh,.This positive development came about both on account of the goodwill of the caretaker administration, but also, most importantly, on account of a judgment of the Supreme Court of Bangladesh (High Court Division) in a Public Interest writ petition filed in 2008, which instructed the government to put effect to this law, which was passed way back in 2003.

CONSTITUTIONAL PROVISIONS AND INDIGENOUS PEOPLE IN BANGLADESH

Constitutional Provisions during British & Pakistani Rule

Most of plains Bangladesh became a part of the British East India Company's territory by the mid-1760s. The CHT, however, was not annexed to Bengal until 1860. It became a British tributary after the defeat of the Chakma king at the hands of the Company's forces, in 1787, ending a decade-long guerrilla war. Since the early part of British rule in South Asia, areas inhabited by indigenous people were administered under special constitutional dispensations. Laws applicable to other areas were considered to be too complex to be suitable for these areas, and hence they were generally administered through special regulations that left the day-to-day administration in the hands of traditional chiefs and sub-chiefs, with supervisory authority being reserved for the colonial authorities, and exercised through residents, agents, superintendents or district officers (deputy commissioners). In-migration of non-indigenous people, and trading and money-lending by non-indigenous people, were strictly regulated. These special laws included the Inner Line Regulation, 1873, the Scheduled Districts Act, 1874, the Chin Hills Regulation, 1896 (Regulation V of 1896), the Government of India Act, 1915, and lastly, the Government of India Act, 1935.

Under the Government of India Act, 1915 (as amended in 1919), the CHT, like other similar areas in British India, was regarded as a "backward tract". This designation was changed to "excluded area" in the Government of India Act, 1935. The 1935 Act functioned as the provisional constitution of Pakistan until its first constitution was adopted in 1956. The 1956 Constitution of Pakistan retained the 'excluded area' status of the CHT and other such areas, but changed the denomination to "tribal area" in the following constitution, known as the Pakistan Constitution of

1962. However, through a constitutional amendment in 1963 – through the Constitution (First Amendment) Act, 1963 – the name of the CHT was excluded from the list of tribal areas. Since then the special constitutional status of the CHT has never been revived. One of the direct impacts of this was the striking down, in 1964, of a CHT law, that was found by the High Court of East

Pakistan to be contrary to the freedom of movement clause of the then constitution. More recently, the constitutionality of the CHT Regional Council Act of 1998 and the Hill District Councils laws of 1998, and even the constitutionality of the CHT Accord of 1997 itself, have been challenged in the High Court, and the hearings on the matter are yet to be concluded.

An 'excluded area' (and later, a tribal area) denoted an almost exclusively tribal-inhabited area, where general laws and regulations were very sparingly applied, and whose administrative features were unique to the territory. A 'partially excluded area', on the other hand, denoted a comparatively mixed, but still predominantly tribal-inhabited territory, whose administrative system was more integrated into the regular administration of the province (than was the case with excluded areas), and a higher number of laws of general application were allowed to function therein.

Under the Government of India Act, 1935 and the Pakistan Constitution of 1956, the CHT formed part of an excluded area, while the areas of Sribardi, Nalitabari, Haluaghat, Phulpur, Durgapur and Kalmakanda in the greater Mymensingh region were recognized as 'partially excluded areas'. However, when the former excluded areas were re-designated as 'tribal areas' in the 1962 Constitution of Pakistan, the former partially excluded areas of Mymensingh were not included in the tribal areas list, unlike the CHT (but which too lost this status in 1964).

In India, most of the former excluded areas – for example in parts of Northeast India, such as in Arunachal Pradesh, Meghalaya and Mizoram-became autonomous district or regional council areas with legislative authority under the 6 Schedule to the Constitution of India (within their respective states, now controlled by the indigenous people) while most partially excluded areas – such as in Jharkhand and Chattisgarh states and in other parts of peninsular India-became specially administered areas under the 5 Schedule, where legislative authority is vested upon the provincial governor with the assent of the President.

The provisions of the 5 Schedule areas are compatible with the provisions of Conventions No. 107 (especially article 27) and 169 (especially article 33), but some of its provisions, and particularly, several provisions of the 6 Schedule-which recognize the wide-ranging autonomous authority of district and regional councils over legislation, administration of justice, land and development, among others-go beyond the ILO Conventions No. 107 and 169, and are closer to those of the UN Declaration on the Rights of Indigenous Peoples. The relevant provisions of the Declaration include those on *self determination, autonomy* and the *maintenance of indigenous peoples' juridical and other institutions* (respectively articles 3, 4 and 5, and 34), *participation* (especially articles 18 and 19), *development* (articles 20, 21, 23) and *land* (articles 25-30), among others.

The 1972 Constitution & Demands for Constitutional Recognition

At the time of Bangladesh's independence from Pakistan in 1971, the excluded area status of the CHT and the partially excluded area status of indigenous-inhabited parts of greater Mymensingh had not been revived. In 1972, before the national constitution was adopted, the late leader of the JSS and Member of Parliament from the northern CHT, M. N. Larma, demanded the revival of the special constitutional status of the CHT and formal constitutional recognition of the hill peoples and other indigenous peoples. The demand was rejected and Larma was advised to join the mainstream of Bengali nationalism.

Since then, again from 1992 onwards, when the *International Year of the World's Indigenous People* was celebrated in different parts of the world, and in Bangladesh, more and more voices could be heard demanding the formal recognition of the Adivasi and hill peoples in the national constitution. During the time of the eighth parliament led by the BNP, the Member of Parliament from Rangamati district, Mani Swapan Dewan, is known to have raised the issue of constitutional recognition in parliament, after a gap of more than thirty years.

Now, with the landslide victory of the Awami League-led *Grand Alliance,* which has the requisite two-thirds majority to make constitutional amendments, the issue of constitutional reforms becomes alive again.. At a public meeting in Dhaka in January 2009, the CHT Affairs State Minister Dipankar Talukdar, MP pledged his support for constitutional recognition of the formal status of indigenous peoples as "Adivasi".

Implementing the Equal Rights & Anti-Discrimination Clauses of the Constitution

Despite the fact that there is no direct mention of the indigenous peoples of Bangladesh in the national constitution, there are a number of provisions in it that are of direct relevance to their rights. Many of these provisions are of concern to the indigenous peoples on account of their situation of disadvantage, under the generic description of "backward section of citizens". The reference to the *backward section of citizens* occurs both in Part II-which declares the *Fundamental Principles of State Policy* (particularly, article 14, and also, article 19)-and in Part III-which lays down the *Fundamental Rights* (particularly, articles 27, 28 and 29). These provisions have three distinct, but related, implications, among others, for the rights of indigenous peoples. Firstly, they totally outlaw discrimination based on race, religion etc. (a situation that indigenous peoples in Bangladesh, as elsewhere, have continually suffered from), drawing upon anti-discrimination provisions of the *Universal Declaration of Human Rights*. Secondly, they urge the Bangladeshi state to take *affirmative action* for the welfare of its disadvantaged groups. Thirdly, these provisions protect those affirmative acts of the state that might otherwise amount to discrimination. These provisions are clearly of the nature that are contemplated by article 3 of Convention No.107, which concerns special protective measures in order to ensure that the disadvantaged situation of indigenous and tribal people does not prevent them from enjoying the benefits of the general laws of the country. Article 3 also clarifies that these protective measures should not prejudice the enjoyment of general rights of citizenship that are available to indigenous and tribal people as citizens of the country concerned.

There is no doubt that the direct, respectful and unequivocal recognition of the status and cultural identity and integrity of all the peoples of Bangladesh is important, not the least, to correctly reflect the country's multicultural composition and heritage in the constitution. It is to be hoped that the situation will be corrected through amendment in future. However, further delays in implementing existing constitutional provisions in the hope of improving upon existing constitutional provisions has the risk of subjecting the already disempowered and marginalized indigenous peoples of the country to further exclusion beyond a point after which remedial action will become meaningless. It is of the utmost importance therefore, to proactively promote the implementation of the existing provisions of the constitution that favour indigenous peoples' rights, rather than concentrate solely upon achieving some sort of formal constitutional recognition of the status of the indigenous peoples. The provisions on equal opportunity in public employment (Article 29(4), for example, provide a clear case of under-implementation, if not non-implementation. Administrative reforms to enable indigenous peoples to substantively participate in decision-making processes on state employment matters – such as in the *Public Service Commission*-could help remedy the situation to a great extent. Many other such steps could be taken to advance affirmative action programmes for situations already accounted for in the present constitution, and strongly supported, in the case of the CHT, by its special legal and administrative system. The aforesaid constitutional provision is compatible with article 15 of Convention No. 107 and several of the provisions of ILO Convention No. 111.

OVERVIEW OF GOVERNMENT POLICY ON INDIGENOUS PEOPLES

This report focuses primarily on Bangladeshi laws vis-à-vis the provisions of the ILO Convention No. 107. However, laws and policies generally go together, and in order to understand the implications of various laws, it may be necessary to also look at the related formal and informal government policies on indigenous issues. This chapter therefore provides a bird's eye view of government policy on indigenous issues.

The Government of Bangladesh has no formal and detailed policy as regards the indigenous peoples of the country in any single document. This is a result, primarily, of the longstanding tradition of the major Bangladeshi political parties not having detailed policies on indigenous issues themselves. Thus the government's policy towards its indigenous peoples has to be gleaned from laws and practices followed by successive governments and by identifying provisions of the sectoral policies – e.g., on Education, Health, etc. – and Poverty Reduction Strategy Papers, where they deal with indigenous peoples expressly or on account of default, i.e., policy neglect in not having express provisions where such are relevant.. The most important of the applicable constitutional and other laws have been referred to in the preceding chapters (chapters III, IV and V). This chapter will provide a broad overview of important elements of the government's policy towards the indigenous peoples, (i) by analyzing the impact of the government's major programmes on governance, development and resource rights issues in the plains regions; (ii) by looking at the state of implementation of the CHT Accord of 1997; (iii) by an overview of major sectoral policies; and (iv) by attempting to understand the implications of the *Poverty Reduction Strategy Papers* ("PRSP-I" and "PRSP II").

It is important that the development policy of the Government of Bangladesh concerning its indigenous population is understood within the context of the government's responsibilities under Convention 107. In other words, where such policy falls below the standards mentioned in the aforesaid Convention, it is incumbent upon all concerned to attempt to raise the policy to meet the standards of the Convention. Most of the provisions of the Convention are relevant to the arena of development. In particular, the provisions on *protective governmental action* (article 2), *education* (articles 6, 21-26), *land* (articles 11-14), *recruitment and conditions of employment* (article 15), *vocational training, handicrafts and rural industries* (articles 16, 17, 18) and *social security and health* (articles 19, 20) are directly relevant to account for in adopting and implementing a development policy for indigenous peoples.

Government Programmes affecting Indigenous Peoples in the Plains Regions

Policy neglect concerning land alienation issues is among the areas of the highest concern for indigenous peoples of the plains, including in the case of the "eco park" project in the Mandi-inhabited Madhupur forest area, the "national park" project in the Khasi-inhabited parts of Maulvibazar district, and the eviction of Santal people in Dinajpur district in the name of "social forestry". To this may be added the threatened eviction of indigenous people in Dinajpur district for a coal mining venture of a multinational company. Demands for resolution of the land dispossession problems through a Land Commission for the plains, and for reservation of seats in the national parliament and in local government bodies, has grown in recent years. Another major area of concern has been the continued exclusion of the plains indigenous peoples from decision-making processes of the *Special Affairs Division,* the premier national-level institution that deals with Adivasi issues of the plains. Although the existence of the *Special Affairs Division,* and the broad ambit of the Division's scope of work are consistent with article 27 of Convention No. 107- which concerns the creation or development of development agencies, and relevant legislative and supervisory measures – the absence of participation of representatives of the peoples concerned is contrary to the provision of article 5 of the Convention, which concerns the participatory rights of indigenous people *(participation and collaboration).* Such policy neglect is also contrary to several provisions of the Constitution of Bangladesh, including those on equality and equal opportunity (articles 27, 28). Partly as a solution to the problem of non-participation in the affairs of the aforesaid Division, Adivasi leaders have demanded the establishment of a separate Ministry of Adivasi Affairs. Many of these mobilization efforts have become more organized and united than ever before, but have failed to so far translate into the required governmental action. Nevertheless, the positive feature of these developments is that these issues are becoming mainstreamed, through reasonably generous coverage in the national media, and provide some hope of remedial action in the near future. In the context of the recent parliamentary elections, which brought in four indigenous members and about an equal number of ethnic Bengali members publicly committed to indigenous or Adivasi issues, there is reason for some hope.

The Quota Policy

The Government of Bangladesh has a policy of affirmative action to reserve positions in government service and in a number of medical and engineering institutions for members of "tribes". Similarly, semi-government and autonomous universities and a number of other educational institutions also offer quotas for admission. Thus, 5% of government jobs are to be reserved for tribals. However, in practice the due positions are not offered to indigenous candidates, especially where the number of vacancies are small and the indigenous candidates lose out with the arithmetic of fractions. With regard to high positions in government, 2009 saw a marked improvement with the highest number of indigenous appointees, including in the coveted Ministry of Foreign Affairs. However, whether such a trend can be maintained will depend on whether the system of appointment by the Public Service Commission is adequately monitored to ensure fairness. In comparison, the reservation system for lower grade posts with primacy to tribal candidates in the CHT, particularly in the Hill district councils, works much more efficiently. This also has the formal sanction of the government.

Unlike in the case of jobs, the problems in the quota for admission into tertiary-level educational institutions has been far more transparent and fair, but here the problem is not so much of policy implementation, but the inadequate number of seats so offered. Both in the case of jobs and admission to educational institutions, the most extensive system of affirmative action in the world is perhaps practised in India for "scheduled castes", "scheduled tribes" and "other backward classes", championed in the 1950s by the renowned low caste jurist, Dr. B. R. Ambedkar.

The CHT Accord of 1997 and its Implementation

The CHT Accord of 1997, ended twenty-five years of political unrest and a guerrilla war in the region, and sought, among others, to strengthen the CHT self-government system. The agreement resulted in the passage of a number of laws, including the *CHT Regional Council Act of 1998*, the *Hill District Council (Amendment) Acts of 1998*, the *CHT Land Disputes Resolution Commission Act of 2000* and the *CHT Regulation (Amendment) Act of 2003*.

It also resulted in the establishment of three new bodies: a regional council for the CHT, a separate ministry to deal with affairs of the CHT region, and a *Commission on Land* to resolve land-related disputes. Immediately after its signing, it was hailed nationally and internationally, and the then prime minister (who is also the incumbent prime minister for the second time) even received a peace prize from UNESCO. On the other hand, there were those who condemned the agreement. One of these was a faction of indigenous student and youth groups hitherto allied to the JSS – now known as the United People's Democratic Front

('UPDF') – who rejected the Accord as being too weak on autonomy and constitutional safeguards. On the side of the Bengali community living in the CHT, initially, large sections of its population looked upon the agreement with suspicion, fearing that it would undermine their rights and interests. Gradually, a growing number of Bengalis, particularly descendents of the Chittagonean-speaking agricultural migrants of the 19 century, started to align themselves with indigenous people supporting the agreement.

Recently, in their pre-election manifesto, the UPDF openly expressed its support for the implementation of the Accord. Similarly, the manifesto of the Bangladesh Awami League-to which all three MPs from the CHT belong, and which is the ruling party in government – also supported the implementation of the Accord. These developments show new alignments that favour the peace process that was initiated through the Accord, although many other political and bureaucratic difficulties still stand in the way. In meetings with members of an international human rights watchdog body on the CHT called the CHT Commission, members of the newly Despite the fact that the Accord is now an administrative reality, and 'demonization' campaigns against it have visibly weakened, its follow-up process was seen to run into several difficulties, especially during BNP-led rule from 2001 to 2006, when the process of devolution of authority to CHT institutions was almost halted.

Among the major provisions of the Accord that remain unimplemented or under-implemented are the dismantling of temporary military and para-military camps, rehabilitation of the internally displaced, resolution of land disputes, and the transfer of law and order, and authority to recruit local police, to the hill district councils. Although the constituent parties of the former BNP-led coalition government had opposed the accord when they were in opposition, and had even threatened to repudiate the agreement, once they were part of the government, they were forced to deal with the institutionalized creations of the accord (e.g., the CHT Regional Council and the Ministry of CHT Affairs). For example, in May, 2006, the BNP-led government transferred authority over Youth, Textile and Vocational Training institutes to the hill district councils. Therefore, although a detailed scrutiny of the agreement shows large gaps in implementation, especially in the period when the BNP-led alliance was in power, it can be said that on the whole, the accord has achieved a level political, if not constitutional, entrenchment, by default, if not by design.

The National Poverty Reduction Strategy Paper (PRSP-I and PRSP-II)

The PRSP – I

Published in October, 2005, the first *National Poverty Reduction Strategy Paper* of Bangladesh ("PRSP-I"), displays, in contrast to other developments in recent years, a respectful approach towards the rights and welfare of indigenous peoples. The term used to refer to the indigenous peoples is "Adivasi/Ethnic Minority", which is far more acceptable to the people concerned, than 'tribal' or 'upajati'. The document acknowledges the indigenous peoples' history of exclusion and experience of discrimination, among others.

The PRSP-I refers to the "inadequate representation [of Adivasi/Ethnic Minorities] at various levels of government and policy processes", hampering their possibility of influencing policy decisions that affect their lives. In addition, it acknowledges their comparatively low opportunities in education (especially in remote areas), and their difficulties in accessing necessary information. Among the "actions to be taken" that it recommends are the full implementation of the CHT Accord, activation of the *CHT Land Commission* and the *Task Force on Refugees*, resolution of land and forest-related problems in the plains (particularly the "Eco Park"), prevention of 'land grabbing' and 'displacement', increasing access to education, including in the mother tongues of the groups concerned, affirmative action for jobs, and the formation of an inclusive advisory body to advise on matters pertaining to Adivasi issues, including at the *Special Affairs Division*, which deals with Adivasi affairs of the plains.

The primary drafts of the PRSP contained hardly any matters of substance concerning indigenous issues. After intense lobbying by indigenous activists, the cooperation of the consultants concerned, and the cooperation of senior officials of the Planning Commission and the Prime Minister's Office, extensive consultations were held with indigenous leaders from the plains regions and the CHT. Many of the recommendations of these leaders were accepted. The proposals that were not accepted included the identification of the Ministry of CHT Affairs and the Special Affairs Division as focal institutions, and a number of other matters that would have provided more details in the matrices section.

It was hoped that this would help integrate and mainstream Adivasi issues into the general development programmes of the Government of Bangladesh. The proposals also demanded a role for indigenous peoples in implementing and monitoring the document. The acceptance of the aforesaid recommendations would have strengthened the document and its operational context, but the PRSP-I as it stood then was nevertheless an important tool to further the rights of the indigenous peoples of Bangladesh. The latter part of the consultation process on the PRSP was also an example of a positive and reasonably fruitful dialogue between indigenous leaders and government officials. It was hoped that this would show the way towards other cooperative indigenous-government dialogues in the near future. This would also be in accordance with the fundamental concepts of 'participation' and 'consultation' as a cornerstone of ILO policy on indigenous peoples, whose basis was provided in article 5 of Convention 107, supplemented with recommendation no. 37 of ILO Recommendation No. 104, and elaborated further in articles 6, 7 and 33 of Convention No. 169, and several recent resolutions of the International Labour Conferences.

The PRSP – II

Although PRSP-I was a landmark improvement in government policy as regards indigenous issues, PRSP-II provides further entrenchment of the importance of indigenous issues in governance and poverty reduction. In the first place, the phrase "indigenous people" is used interchangeably with "indigenous communities" (PRSP-I referred to "Adivasi/Ethnic Minority"). Secondly, PRSP-II includes a vision statement that acknowledges the preservation of the "social and cultural identity" of the indigenous peoples, and the need to ensure the exercise of the indigenous peoples' "social, political and economic rights" and their "security and fundamental human rights". Thirdly, in the section on future actions to be undertaken by the Government, in addition to reiterating the importance of implementing the unimplemented provisions of the CHT Accord of 1997 (the PRSP-I referred to "full implementation" of the Accord), PRSP-II mentions the ratification of ILO Convention No. 169 and the implementation of the provisions of the UN Declaration on the Rights of Indigenous Peoples. Other important matters for action in include land rights, participation in development programmes affecting indigenous peoples, human development, empowerment of indigenous communities, indigenous peoples' languages and children's access to education, access to electricity, and the mainstreaming of indigenous issues in national policies, including for the key ministries of Land, Agriculture, Education, Primary and Mass Education, Health, Local Government, Rural Development & Cooperatives and Social Welfare. In the section on *Challenges*, the document acknowledged the absence of census and other statistical data on indigenous peoples.

The process of consultation in the case of PRSP-II was arguably far less inclusive than was the case with PRSP-I. However, what is of importance for the indigenous peoples of Bangladesh is that one of the three formal consultations were held in Rangamati, the headquarters of the CHT, and that the word indigenous people has found its way into a formal government document again. Another crucial difference between PRSP-I and PRSP-II as it concerns indigenous peoples is that the provisions of PRSP-II on indigenous peoples are now anchored to two identifiable government institutions, the Ministry of CHT Affairs for the CHT and the Special Affairs Division for the plains, and moreover, the two institutions' activities are sought to be further mainstreamed into existing and future development programmes of the government, including through other key line ministries. This provides a stronger link to adequate budgetary allocations. However, PRSP-II's biggest weakness is perhaps the absence of express commitments to make the Special Affairs Division of the Prime Minister's Office – which deals with budgetary, developmental and other matters of indigenous issues pertaining to the plains districts –, more inclusive and participatory for indigenous peoples of the plains, since they have no direct participation at decision-making levels of the Division. This will hopefully be addressed by the newly-elected government.

A GENDER & EQUITY AUDIT OF BANGLADESHI LAWS & POLICIES ON INDIGENOUS PEOPLES

Equity & Gender

The constitution of Bangladesh contains several provisions that seek to eradicate discrimination against women.Moreover, Bangladesh has ratified the *Convention on the Eradication of Discrimination against Women* (CEDAW), although it has put some reservations, including on the issue of personal laws. Bangladesh has even established a Ministry of Women's Affairs, but the actual situation of women in the country is far from good. Gender-based discrimination is a common phenomenon in Bangladeshi society, including within the indigenous population of the country (as in various other parts of the world). It has been said that indigenous women are even more marginalized than their non-indigenous counterparts, on account of their multiple situations of disadvantage, including "gender, ethnicity and class". Generally, indigenous women in Bangladesh have more social mobility than non-indigenous women, but in terms of inheritance rights and political rights, among others, their situation is said to be little better than that of non-indigenous women, if at all. The problem is compounded by the fact that the prescriptions generally provided or offered by state and civil society actors are based upon "atypical cases of discrimination faced by Bengali women in the plains regions ....[and hence] not appropriate for preventing discrimination against indigenous women". It has therefore been said that these mainstream concepts on remedial measures – including on an ungendered Family Code and use of special ('Family') courts-need to be amended to fit the situation of Bangladeshi indigenous women from the different ethnic and religious groups.

Women are severely under-represented in the traditional and largely hereditary institutions of the CHT chiefs (*rajas*), headmen and karbaries. In the plains too, for example, among the Santal and the Oraon, clan leadership is restricted to men. Even among the matrilineal Garo and Khasi peoples, traditional leadership offices of *Songni Nokma* and *Myntri*, respectively, are generally held by men. In the case of elective offices as well, representation of women is inadequate. For example, only two out of the 16 tribal seats of the CHT Regional Council, and only six out of the 20 tribal seats of each hill district council (when elected councils eventually sit), are reserved for tribal women. Discontent has been particularly high in the case of the interim district councils, which have five members each (including the chairperson), although they are to have a total of 31 members each (including the chairperson) when elections are ultimately held. At the moment, none of the five members of each of the three councils is a woman. Indigenous women have demanded that this situation be corrected by amendments of the law, and have similarly demanded adequate allocation of seats for indigenous women for the reserved women's seats in the national parliament.

The situation of indigenous women in the customary personal law systems shows both fair treatment and discriminatory treatment. However, violence in domestic situations is not uncommon. Moreover, discriminatory practices in inheritance, marriage and child custody still continue. Indigenous women in the urban areas of the CHT are becoming increasingly vocal about demands to outlaw polygyny and to achieve equal inheritance and child custody rights. The situation in the plains regions is also a case for concern, and warrants close attention, including strengthening of women's rights networking alliances.

Over the years, more and more indigenous people, particularly women, have started working in cities. The most common occupations are in beauty salons/parlours (mostly Mandi/Garo) and in factories (particularly Chakma). A study on Mandi women in beauty parlours of Dhaka city shows how vulnerable these women are to exploitation. Legal loopholes make them even more vulnerable since such work is considered to be outside the formal sectors, and therefore outside the full protection of the law. No research is known to have been done on the indigenous women working in garments factories in Chittagong and Dhaka cities. There is little doubt that measures need to be taken to ensure that their rights under Bangladeshi law, and under the Convention No. 107, are adequately protected. Some of these pertinent issues were raised by the *Indigenous Women's Forum* at the Beijing Conference of 1995.

Corrective legislative and administrative measures would be in tune with article 15 of Convention No.107, which seeks, among others, to prevent discrimination between indigenous and other workers. Absence of necessary protective measures for indigenous women workers, therefore amounts to substantive, if not procedural, inequality. This is also contrary to the provisions of CEDAW, and provisions of the Constitution of Bangladesh on grounds of discrimination based on gender.

Equity & Ethnicity

Despite sharing the same generic identity of indigenous, Adivasi, 'tribal' or otherwise, the situation of the different indigenous peoples in terms of political, economic, and social empowerment, or disempowerment, as the case might be, is varied and unequal. The groups with the smaller populations generally tend to be more politically marginalized than the more numerous groups, since their electoral strength is too small to enable them to get their own people elected to office. This is so both in the plains and in the CHT. In the case of the plains, only a few of the aboriginal peoples have been recognized through the East Bengal State Acquisition & Tenancy Act, 1950. Hence those that have not been so recognized are liable to be deprived of the protection offered by the 1950 land law. This may also mean that they will be deprived of economic programmes through the *Special Affairs Division*, which is responsible for overseeing development programmes for the plains indigenous people. There are other instances of discrimination, and exclusion of Adivasis from their due recognition, as mentioned earlier in section 2.2 above.

In the case of the CHT, official recognition is not a problem for the indigenous peoples that were previously recognized through the CHT Regulation, 1900. However, there is discontent among *some* of the indigenous peoples with small populations because they have not been provided direct representation in the CHT Regional Council. In the case of the hill district councils too, representation of the peoples with small populations has in some instances been indirect or otherwise inadequate. This situation needs to be addressed in an equitable manner. For the time being, the regional and district councils could bring about a consultative process to ensure that the representatives of all the indigenous peoples and communities are able to have a say in the major decision-making processes affecting them. This is also important because all the indigenous peoples are not adequately represented in the 'traditional' system of raja or chief, headmen and karbari either. Any such measure, as suggested above, would be consistent with provisions of Convention No. 107 and the Constitution of Bangladesh.

As in the case of the less numerous peoples in the CHT, the self-government systems of the plains indigenous peoples have been de-recognized long since. This situation needs to be remedied, for example, by amending the elective local government body system to create 'reserved' seats in appropriate areas. Moreover, the formerly prevailing system of retaining Welfare Officers at district level – preferably from local indigenous people – could be revived or introduced, as the case might, in the original or amended form, as appropriate. Demands to establish a separate Ministry of Adivasi Affairs, as demanded by Adivasis, could also improve the situation in the plains. Such reforms would be consistent with articles 7, 8 and 10 of Convention No. 107, which provides safeguards for the customary laws, cultural heritage and integrity of indigenous peoples (article 7, 8, 10).

As in the case of political representation, the situation of social and economic opportunities among indigenous peoples is also unequal. Access to education among some of the indigenous peoples – e.g., in the northwestern Rajshahi division, and among the least numerous peoples in the CHT – is seriously inadequate, being far below the national literacy rate (which itself is low), and therefore calls for special attention. Indigenous communities inhabiting remote parts of the CHT-including those that have been internally displaced during the conflict, and who still remain un-rehabilitated-are deprived of both education and healthcare facilities. These too require special attention, which will not be forthcoming unless the concerned laws are supplemented with detailed subsidiary rules, regulations, and guidelines providing special measures to ensure true equal opportunity as provided in the Constitution of Bangladesh, the ILO Convention No. 107, and other international human rights treaties applicable to Bangladesh. Concerted efforts to operationalize some of the areas outlined in the PRSP-II could visibly improve the situation of economic and social rights, at the least. In particular, special measures need to be taken to provide state-funded healthcare and primary education facilities in remote and semi-remote areas that do not otherwise satisfy the general criteria on population, location etc. as mentioned in the national health and primary education policies. The aforesaid policies-the National Health Policy and the National Primary Education Policy-require reforms to accommodate the situation of indigenous peoples, and be supplemented with appropriate policy guidelines and administrative action. With regard to the policy on Primary Education, some progress was made during the period of the Caretaker Government with regard to affirmative action provisions on registration of schools, appointment of teachers and so forth.

INDIGENOUS PEOPLES' POLICIES OF BILATERAL & MULTILATERAL DEVELOPMENT AGENCIES

Bilateral and multilateral development agencies play an important role in supporting development programmes and projects in Bangladesh through grants and loans. Many of them also have their own indigenous peoples' policies to guide the context of development cooperation involving indigenous peoples and to enhance the participation of indigenous peoples' in the process of development. Most of them expressly account for the marginalized situation of indigenous peoples. These include those of European Community member states, including Denmark, Germany, the Netherlands, Spain and the United Kingdom. The policies of the Netherlands and Denmark in particular, have a strong rights focus. Most make specific references to international human rights instruments, including ILO Convention No. 169 and the (then draft) UN Declaration on the Rights of Indigenous Peoples, and other relevant international instruments relevant to environment and development, including Agenda 21, adopted at the Rio Earth Summit in 1992, and the Convention on Biological Diversity. The provisions of many of these policies have been influenced by Convention No. 169. Some of the Indigenous Peoples' policies of multilateral lending agencies such as the World Bank, and regional development banks, such as the Asian Development Bank, are generally regarded as *safeguard* policies, in that they aim, among others, to mitigate or soften the negative impact of development interventions. All of the above reflect the adoption of an increasingly co-ordinated approach and common understanding of the problems and aspirations of indigenous peoples among development-related donor institutions, lending agencies and other major development actors in general. The aforesaid perspectives are wholly compatible with the concept of participatory development as mentioned in article 5 of ILO Convention No. 107, which are essentially similar to those of the corresponding sections of ILO Convention No. 169 (particularly articles 6 & 7), although the latter are worded in stronger language, reflecting the progressive development of international human rights law, including the principle of free, prior and informed consent. No doubt, the development-related provisions of Convention No. 107 also need to be understood in the context of the aforesaid progressive concepts and paradigm shifts.

Among these, the most progressive is that of the Government of Denmark. Known as the *Danish Strategy for Indigenous Peoples.* This policy was reviewed by a team of indigenous experts from different parts of the world in 2000-2001, at the invitation of the Danish government.

The team praised its contents but advised better implementation through mainstreaming. Among other relevant policies is the European Commission's policy on indigenous peoples. The aforesaid policies stand out among other comparable documents as being of a 'political' nature, in that they may be invoked to urge the Danish Government and the European Community-major development partners of the Government of Bangladesh-to not only follow the policy in their development projects, but to raise the issue of the indigenous peoples' rights in their dialogues and interface with national governments, and inter-governmental and UN agencies. At the international levels, and in the countries of its bilateral cooperation, including in Bangladesh, the Government of Denmark has been actively supportive of the rights of indigenous peoples.

The UNDP also has its own policy on indigenous peoples. In 2003, on the basis of complaints that various components of the UNDP's project entitled *Promotion of Development and Confidence Building in the CHT* were contrary to the UNDP's own policy on indigenous peoples (*UNDP and Indigenous Peoples: A Policy of Engagement*) a team was sent by UNDP headquarters in New York to the CHT in January, 2004 to look into the matter. Eventually, the dispute was resolved amicably. Thus this was an example of advocacy work that successfully invoked an Indigenous Peoples' policy of a development institution in defence of indigenous peoples' rights.

The experiences of indigenous peoples of projects funded by the World Bank and regional development banks (including the Asian Development Bank) in different parts of the world has been far from happy. The support of the Asian Development Bank in the forestry sector in Bangladesh is regarded as having had negative impacts on the rights of indigenous peoples. A *potentially* positive example of the Asian Development Bank's involvement in a project involving indigenous peoples in Bangladesh is the *Chittagong Hill Tracts Rural Development Project*. The project is known to have been initiated with the consent of the Chittagong Hill Tracts Regional Council, the premier institution of the Chittagong Hill Tracts region, and includes the indigenous chairman of the council as the *ex-officio* chairperson of the project's *Regional Coordination Committee*. However, there is a potentially negative aspect of the project in that a component on micro-credit (small loans) appears to have been incorporated into the project at the behest of the recommendations of the bank's president, rather than the wishes of the indigenous peoples of the region. In any event, one may consider where there it is still too early on in the project cycle to judge the success or failure of the project with regard to the rights and welfare of the indigenous peoples of the CHT.

The European Commission (EC) – which also has its own Indigenous Peoples policy-is also a major development partner of the Government of Bangladesh, and one of the major funders of a UNDP-managed project in the CHT. It has also funded development projects in various other countries of the world, including in areas inhabited by indigenous peoples. In 2003, the EC had some of its projects in indigenous areas evaluated by indigenous experts and the findings shared among indigenous peoples, EC project staff and policy-level EC officials. Similar undertakings could be taken for projects funded by the Asian Development Bank and other such funders of projects in Bangladesh. Similarly, indigenous peoples have demanded that the World Bank should involve indigenous organizations in actively tracking and monitoring the bank's operations through the whole project cycle. Recently, the World Bank replaced its existing *Indigenous Peoples* policy, OD 4.20, with a new policy, known in short as OP 4.10/BP 4.10. Although the bank included many seemingly beneficial provisions that are respectful towards the rights of indigenous peoples, most of these provisions appear to have more of a recommendatory, than a mandatory, character. The Asian Development Bank is reviewing its existing policy on Indigenous Peoples. It is to be hoped that the experience of the World Bank will not be repeated in this case, but the prospects may not be all that bright.

INTERNATIONAL HUMAN RIGHTS LAW & THE INDIGENOUS PEOPLES OF BANGLADESH

Internationally, and especially within the United Nations (UN) system, high focus has been put on indigenous issues over the last decade or so. Landmark decisions and programmes within the UN include the declaration of 1993 as the *International Year of the World's Indigenous People,* the declaration of the decade between 1995-2004 as an *International Decade for the World's Indigenous People,* the appointment of a *Special Rapporteur on the Human Rights and Fundamental Freedoms of Indigenous Peoples* in 2001, the establishment of the *UN Permanent Forum on Indigenous Issues* in 2000 the adoption of the UN Declaration on the Rights of Indigenous Peoples by the UN General Assembly in September 2007 and the establishment of the *UN Expert Mechanism on the Rights of Indigenous Peoples,* which reports to the Human Rights Council, which held its first meeting in October, 2008.

The adoption of the Declaration on the Rights of Indigenous Peoples came about through the commitment of the international community as expressed in several global meetings, including the 2005 World Summit. The Declaration is hoped to draw much-needed attention towards the situation of indigenous peoples worldwide, and to encourage legal and administrative reforms, in a spirit of partnership with indigenous peoples. The establishment of the Expert Mechanism in 2007 – whose members are all experts of indigenous origin from different parts of the world-is another major development for the promotion and protection of the human rights of indigenous peoples. This body is expected to take over some of the functions earlier fulfilled by two subsidiary bodies of the Commission on Human Rights, namely, the UN Working Group on Indigenous Populations and the UN Sub-Commission on the

Protection and Promotion of Human Rights. However, since this body will report directly to Human Rights Council, one could expect that the human rights of indigenous peoples will now receive attention at a higher level within the United Nations, than before. However, despite having three separate United Nations mechanisms with specific mandates on indigenous issues (the Permanent Forum, the Expert Mechanism and the Special Rapporteur), there are several limitations in the functions of the three mechanisms, as was discussed in-depth at an international seminar held in Madrid in February, 2009.

The Declaration on the Rights of Indigenous Peoples acknowledges several key rights that apply to indigenous peoples, including on self determination, autonomy, implementation of treaties and agreements concerning indigenous peoples, formal acknowledgment of their collective and individual land rights, maintaining indigenous justice systems and customary law, just and prompt resolution of land and other disputes involving indigenous peoples, freedom from militarization, cultural and language matters, and above all, freedom from all forms of discrimination. On the adoption of the Declaration, the most important challenges – apart from converting it into a UN treaty (convention or covenant or as otherwise named) at some point in the future – will be to seek ways to implement it within the UN and other inter-governmental contexts, and through advocacy at country levels. An immediate need is to translate its provisions into as many relevant languages as possible.

Declarations are extremely important in guiding governments, the UN system, and other international fora in taking necessary measures. Such instruments have great moral authority, although from a purely legalistic perspective of state agencies involved in administration, they may be regarded as not "legally binding". Ratified multilateral treaties (whether called covenants or conventions or otherwise), unlike Declarations, are directly binding upon the ratifying countries, which are then subject to the respective monitoring systems, usually entailing reporting responsibilities, among others. However, as far as UN agencies are concerned, the difference between a treaty and a declaration may in many cases be almost irrelevant, at least as far as their own programmes are concerned, as long as these do not violate national or international law. In either case, the agency needs to respect the provisions of these universal human rights standards. In the case of the Declaration on the Rights of Indigenous Peoples, the instrument actually calls upon UN agencies to promote and implement the provisions of the Declaration.Therefore, just as the integrationist aspects of ILO Convention 107 are no longer regarded as relevant, similarly, the provisions of ILO Convention 169 (and 107) need to be looked at anew in light of the positive developments in international law, including the UN Declaration on the Rights of Indigenous Peoples. It is important to underscore that the Declaration does not *create* new rights, but merely reinterprets existing international law through a truly non-discriminatory and progressive perspective.

Among the multilateral human rights treaties ratified by Bangladesh, the ILO Convention No. 107 is of the utmost importance to the indigenous peoples of Bangladesh, since it is the only treaty dealing directly and substantively with indigenous peoples' rights. Other treaties of relevance for Bangladeshi indigenous peoples include the *International Convention on the Elimination of Racial Discrimination* (CERD), the *International Covenant on Civil and Political Rights* (ICCPR) and the *International Covenant on Economic, Social and Cultural Rights* (ICESCR), the *Convention on the Elimination of Discrimination against Women (CEDAW) and the Convention on the Rights of the Child (CRC)*.. In addition, there is the *Convention on Biological Diversity,* which, although dealing primarily with biological diversity, also contains a number of provisions relevant to the resource rights and cultural heritage of indigenous peoples.

Indigenous rights activists from Bangladesh have participated in many international processes over the last decade or so, in the UN Permanent Forum on Indigenous Issues, in a number of subsidiary bodies of the former Commission on Human Rights dealing with indigenous issues, including the *Working Group on Indigenous Populations and the Working Group on the Draft Declaration on the Rights of Indigenous* Peoples, in the first session of UN Expert Mechanism on the Rights of Indigenous Peoples in October, 2008 and most recently, at the Universal Periodic Review under the Human Rights Council in Geneva, in February 2009. However, their participation in the formal processes of the treaty monitoring bodies, including the process of supervision of ILO Convention No. 107, and their engagement with the Office of the *UN Special Rapporteur on the Human Rights and Fundamental Freedoms of Indigenous Peoples,* has been very limited, at best. Unless this trend changes significantly, the impact of these processes on the ground will remain limited. Given their political marginalization in the Bangladeshi polity, it would not be strategic for indigenous peoples to neglect the human rights treaty processes, including the ILO 107 process.

Of course, international human rights processes do have shortcomings. It is sometimes said that they have little or no "teeth" in terms of direct implementation. Thus, these processes can have a persuasive leverage, but no direct "sting" to enforce decisions on any violation of the international laws concerned. In *dualistic* legal systems such as Bangladesh, rights based on international treaties are not directly enforceable in the domestic courts, unlike in *monistic* systems like the USA, Nepal, Netherlands, Spain and several countries in Latin America, where treaty law is considered to be a part of domestic law, and hence directly enforceable in domestic courts. Despite these limitations, there is little doubt that participation in the aforesaid processes does lead to pressure upon the government to show adherence to its international human rights obligations, even though the exact extent of such impact is difficult to gauge. Another difficulty in this respect is that such pressure does keep a government on its toes for a short while, but it is sometimes difficult for the activists to sustain such pressure on governments for long, due to constraints of funds, and organizational and networking difficulties. Thus capacity-raising and organizational strengthening of indigenous peoples' organizations and networks remains a high priority for any serious human rights activist.

Interactions between governmental and indigenous participants at international fora can on occasions be quite conflictual, but this need not necessarily be so. Healthy debates between human and indigenous rights activists and government diplomats can and does happen, and these interactions have the potential to improve overall efforts to deal with concerned situation, at least over time, in some cases. Such dialogues should be encouraged, and this would be consistent with the provisions of

Convention No. 107 with regard to *administration* (article 27) and the *elimination of prejudice against indigenous people* (article 25).

Ultimately, the long-term goal of any international human rights treaty, including the ILO Convention No. 107, is to implement the concerned rights in the domestic situation. Therefore, international advocacy and networking needs to be combined with corresponding work at the national and local levels. Of course, these international standards can and should guide legislation and practices within Bangladesh with regard to indigenous peoples' issues, but when it comes to the question of actually implementing the concerned rights, it seems that political lobbying is necessary, a sphere in which indigenous peoples in Bangladesh, as in other parts of the world, have remained quite marginalized, especially on account of majoritarian traditions that go under the name of democracy.Therefore, efforts to strengthen alliances and networks across the political, ethnic and civic spectrums remains a challenge. This is especially difficult in the case of the plains regions where there is no separate regional or local government system to guarantee the indigenous peoples with a formal say in the process of governance and administration.

BANGLADESHI LAWS AND POLICIES & CONVENTION NO. 107: A SUMMARY

Comparison of Convention No. 107 and the Laws of the Plains Regions of Bangladesh

The preceding discussion shows that there are several areas in which Bangladeshi law and policy fall short of the provisions of Convention No. 107, particularly in the plains regions of the country. The non-implementation of the land alienation safeguards contained in the East Bengal State Acquisition and Tenancy Act, 1950 (relevant to article 13 of Convention No. 107), the continuing absence of participation of indigenous peoples in the programmes of the *Special Affairs Division* (relevant particularly to articles 5, 15, 25 and 27 of Convention No. 107), and the violation of the land and resource rights of the people of the Madhupur forest and in Sylhet division (relevant to article 13 of Convention No. 107), are but the most publicly known situations of policy failure and policy neglect.

Other areas where the provisions of Convention No. 107 are unimplemented or under-implemented in the plains include language and cultural rights (especially relevant to articles 7 and 8 of Convention No. 107), healthcare and social welfare, education, training and employment issues (especially relevant to articles 15-26 of Convention No. 107). In substance, it could be said that the provisions of ILO Convention No. 107 have remained very largely unimplemented in the plains regions. The relative absence of issues of the plains indigenous peoples in the reports of the Government of Bangladesh to the Committee of Experts (CEACR), and the equally low level of attention of the Committee's experts on plains indigenous issues is indeed most unfortunate. The absence of procedures allowing a direct role of indigenous peoples in the decision-making processes of the ILO, together with the extreme social and economic marginalization of most indigenous communities in the plains stand as major obstacles towards progress in this context. The apparent lack of attention to the plains areas on the part of the CEACR also reflects lack of information received by them. The ILO has a duty to raise awareness about the reporting process and sensitize governments, employers and workers' organisations in this regard, and this needs to be done by bearing in mind the extreme marginality of some of the indigenous peoples and communities concerned, who all too often continue to remain invisible to, and without a voice in, processes wherein they might otherwise obtain *some* redress for the violation of their rights.

Comparison of Conventions 107, 169 and the Laws of the Chittagong Hill Tracts

In contrast to the plains regions, the CHT region shows, at least on the face of it, instances of *both* non-implementation and under-implementation of the provisions of Convention No. 107, *and* instances where law and policy in the CHT go beyond the provisions of Convention No. 107. With a broad overview, one can perhaps see under-implementation of the provisions of Convention No. 107 with regard to health and social security issues (articles 19, 20), and language and communication matters (especially articles 22, 23), at least from the point of view of actual practice. With regard to education and employment provisions (particularly articles 21-26 and 15, respectively), the situation is perhaps more or less at par with the provisions of the convention. However, with regard to the role of the indigenous peoples in governance, administration and the development processes, the situation in the CHT is perhaps somewhat ahead of the standards of Convention No. 107, and closer to the provisions of the more progressive and stronger Convention No. 169. The matrices contained in the table 3 series mention the most important laws of Bangladesh that are directly relevant to indigenous peoples, and their relative situation in comparison to the provisions of Convention No. 107, and contextually, to Recommendation No. 104, which is related to Convention No. 107, and to Convention No. 169, where relevant.

ILO Convention 107 and its Impact on Bangladeshi Laws & Policies

The situation of implementation of the provisions of ILO Convention No. 107 in Bangladesh is far from satisfactory. The overall situation of participatory governance, development and land rights in the plains regions is very clearly indicative of the very little impact that the provisions of the Convention have had in influencing legal and administrative measures to protect the rights and welfare of the plains indigenous peoples. The only instrument that substantively refers to indigenous peoples' rights in the plains – the East Bengal State Acquisition and Tenancy Act, 1950 – is an Act that has its roots in a 19 century law of the British colonial period, and was clearly not the result of the ratification of the convention by the Government of Bangladesh in 1972, or by its predecessor Government of Pakistan, in 1960.

In comparison, the Chittagong Hill Tracts region shows a larger number of legal and administrative measures that safeguard the rights of the region's indigenous peoples, with some even going beyond the standards of Convention No. 107 onto the higher standards of Convention No. 169. These developments, however, are again, very unlikely to have been due to the direct influence of the provisions of the Conventions No. 107 or 169. It is perhaps purely coincidental that many of the relevant provisions of CHT or national law relate closely with, and are compatible with, various provisions of ILO Conventions No. 107 or 169.

A review of the circumstances and periods in which the major pro-indigenous laws of the CHT were formulated show three major periods. The most progressive period is obviously the period after 1997, in which year the "peace" accord of CHT was signed. Of the laws before 1997, there are some reasonably progressive ones that were passed in 1989 (the Hill District Council Acts), during the period of the armed revolt, and others that were mostly framed during the period of British colonial rule (1860-1947). In all three of the cases, the determining factor was clearly not the provisions of the ILO Convention. In the post-1997 period, the key activating agent is the 1997 Accord. As for the pre-1997 period, the 1989 package was a result of another political agreement with CHT leaders (in 1988), when the armed struggle was still continuing, and the rest were doing the colonial period.

ASSESSING THE MONITORING SYSTEM OF CONVENTION NO. 107

Important Features of the Reports of the Committee of Experts

The reports provided by the Government of Bangladesh ("GOB") to the ILO's *Committee of Experts on the* Application of Conventions and Recommendations ("CEACR" or "Committee"), and the queries and *Direct Requests* of the Committee, and its *Observations* show a number of special features or trends, among others. Firstly, the reports of the GOB are quite often not sent at all or sent late. For example, this was mentioned in the Committee's *Observation* of 1994 (64 session), repeated in 1995 (66 session), in 1997 (68 session) in 2000 (71st session), and again in the Committee's Individual Observation published in 2005. Secondly, the GOB has not responded at all, or adequately enough, to the requests made to it to provide information about tribal people of Bangladesh outside of the CHT (1990: 60 session, 1991: 61 session, 1993: 63 session, 1995: 66 session, 1998: 69 session, and in 2000: 71 session). The Committee even referred to specific incidents: including conflicts between the Garo people with the Forest Department (1991, 1993, and 1995). Thirdly, queries about CHT laws – including on the status of the CHT Regulation, 1900 – seem to have been left largely unanswered (1993, 1994, 1995, and 2005). The aforesaid pattern of reporting indicates, among others, that the GOB has continually been evasive about the status of the CHT Regulation, has largely ignored the situation of the indigenous people in the plains, and most importantly, sent the reports in such a manner, and at such periods, that it enabled it to avoid a focussed and sustained monitoring of particular aspects of its responsibilities under the Convention.

On the other hand, the focus of the Committee too seems to have been sustained on certain issues (e.g., on the CHT Regulation, 1900), but wavered in others – such as on the situation of the plains indigenous people (which showed fewer specific queries), in fact almost mirroring the GOB's example. Fourthly, it is seen that in a limited number of instances, the Committee has referred to observations of other human rights monitoring processes concerning the situation of indigenous people in Bangladesh, such as observations of the *UN Special Rapporteur on Torture and other Cruel, Inhuman or Degrading Treatment* (1995, 1997, 1998, 2000) and some observations of the *UN Committee on the Elimination of Racial Discrimination* (1993, 1994, 2001). Such references no doubt add to the combined impact of human rights monitoring processes upon the situation on the ground, particularly bearing in mind the disadvantages faced by indigenous people and their consequently marginal role in facilitating effective monitoring of formal human rights processes.

Obstacles towards Effective Monitoring: Access to Data & Marginality of Indigenous Peoples

With regard to certain issues-such as on land-related laws of the CHT, and the role of the chiefs in administration-the queries of the CEACR suggest access to reasonably detailed knowledge on the part of its members. In other cases-such as on the status of the CHT Regulation, 1900-, both the questions asked (by CEACR), and the answers provided (by the Government of Bangladesh), suggest absence of clarity in conception. Therefore, effectiveness of the monitoring process may depend on the relative expertise of the 'experts'.

However, the expertise of the CEACR members may also depend upon the nature and extent of access that they have to relevant information. Such information may not be easily available for a number of factors, including the non-cooperation of the government, the dysfunctionalities within the system of governance in the country, and most importantly, the marginalized conditions of the indigenous peoples concerned.

The more marginalized they are, the more likely it is that information about the violation of their rights remains un-communicated at worst, or only locally disseminated at best on account of language and technical factors. It is therefore extremely important that the C.107 monitoring process accounts for two significant features that are unique to the process. Firstly, that this Convention deals with a section of humanity, the world over (including in Bangladesh), that is perhaps second to none in terms of political, social and economic marginality. And secondly, that the process of ILO Conventions 107 and 169 (both are identical) is perhaps the only international multilateral treaty process in which the subjects of the treaty (indigenous and tribal people in this case) do not have a formal standing (*"locus standi")* to either complain directly to the monitoring body (the CEACR), or otherwise participate in other relevant decision-making processes (such as the International Labour Conferences, which are restricted to the 'tripartite' system involving government members of the ILO, employers' associations and trades unions).

At the country level, shortcomings faced by indigenous people in conducting dialogues with their own government, including the focal institutions – in this case the *Ministry of CHT Affairs*, the *Special Affairs Division* and the *Ministry of Labour* – compounds the issue further. Therefore, the combined effect of the disadvantage suffered by indigenous peoples in the aforesaid respects make it all the more difficult for them to help provide necessary information and inputs to enable the members of the CEACR and the government officials concerned to carry out their respective responsibilities under the convention in a more informed manner. Unless and until the aforesaid weaknesses in the chain of supervision and monitoring are addressed in some form or the other, the overall system of monitoring of the Convention is unlikely to improve significantly. *Perhaps what cries out very loud in the present scheme of things is to bring about periodical participatory reviews of the entire ILO system of monitoring of Convention No. 107* (and Convention No. 169*) involving representatives of indigenous peoples, among others.* The review of the *Danish Strategy for Indigenous Peoples* by a team of indigenous experts in 2001, and the experiences of the European Commission's monitoring of projects by indigenous experts in 2003, are examples that can help provide necessary guidance in this respect.

MEETING THE CHALLENGES

Increased awareness about the Convention No. 107, and a marked improvement in the process of implementation of the Convention – including through monitoring by the CEACR – will depend upon the nature of measures that may be taken by the ILO itself (at its head office and in its country offices), by the Government of Bangladesh, and by the indigenous peoples of Bangladesh, who will hopefully act as the main catalyst, since it is their rights that are at stake.

In addition, much may also depend upon what other relevant 'actors' of the process do, including indigenous peoples in other countries that have ratified Convention No. 107, other existing and potential allies, whether they are state members of the ILO, Bangladeshi and foreign employers' associations, Bangladeshi and foreign trades unions, human rights organizations, United Nations bodies and others.

The following are regarded as among the most important areas of focus to help further the rights under this convention, in the context of Bangladesh, and where relevant, to the ILO 107 and 169 processes in other countries.

REVIEW OF ILO 107 MONITORING PROCESS BY INDIGENOUS AND OTHER EXPERTS

A review of the ILO process on Convention No. 107 by experts, including from among indigenous peoples, may provide useful insights into possible measures to help facilitate better monitoring of the implementation of the Convention.

Such a review could be asked to focus on such matters as the dissemination of necessary information, the internal and monitoring processes of the ILO and the nature of indigenous participation therein, the system of governance in the ratifying country and the strengths and weaknesses of courts of law and other justice delivery systems, the presence of human rights commissions and civil society human rights watch dog bodies, the level of sensitization among relevant government leaders and officials, the organizational strengths and weaknesses of indigenous peoples and their organizations and institutions, and the possibilities of increasing synergy through coordination and networking, among others. Some of these are elaborated below.

INFORMATION

Information on the Convention, its formal monitoring process, the tripartite ILO system, and best practices on the implementation of ILO Conventions 169 and 107, among others, need to be widely disseminated, in appropriate language and form. The ILO country offices can play the role of a focal group, while partnering with indigenous peoples' organizations, relevant government agencies and others in sharing responsibilities.

PROCESS AND PARTICIPATION

The existing difficulties of direct participation of indigenous peoples within the ILO system need to be addressed. The establishment of the high-level United Nations Permanent Forum on Indigenous Issues should help encourage the ILO system to itself reform, if so deemed necessary by all concerned. Until such time as reforms to enable the full participation of indigenous peoples in the ILO system are in place, *ad hoc* measures could be taken to provide 'observer status' to indigenous peoples' organizations to enable them to participate in relevant ILO meetings and processes. The experience of indigenous participation during the drafting of Convention No. 169 needs to be revisited to provide practical ideas in this regard.

MONITORING

The formal standing *(locus standi)* of individuals, institutions and organizations of indigenous peoples should be acknowledged to enable them to directly address the CEACR, in appropriate ways (e.g., by communicating written complaints in a fixed format set by the CEACR).

SENSITIZATION

Appropriate measures need to be taken to sensitize and orient members of the CEACR, concerned ILO staff at ILO headquarters and in country offices, and concerned Government Departments, on indigenous issues in general, and on other relevant matters, as appropriate. In the case of the specialized ILO processes, steps may be taken to orient relevant government and ILO staff, who may be more familiar with 'labour' issues than indigenous issues. Parliamentarians, judges and officials of the concerned ministries (such as the Ministry of CHT Affairs, the Special Affairs Division and the Ministry of Labour in Bangladesh) should be included. Whenever international issues of Convention No. 107 arise, the Ministry of Labour should involve other relevant bodies (such as the Ministry of CHT Affairs, and the Special Affairs Division, and relevant self-government institutions of indigenous peoples, such as the regional and district councils and the traditional institutions in Bangladesh). In areas with large concentrations of indigenous people, it should be compulsory for government officials posted therein to be sensitized through training and orientation programmes. Indigenous peoples can streamline their efforts to resist 'demonization' campaigns or misinformation (intended or otherwise) that tend to lead to the portrayal of indigenous peoples' rights as being prejudicial to the rights of others, which it is not. Relevant government bodies, Indigenous peoples' institutions and organizations should take the lead in such matters.

CAPACITY-RAISING

Capacity-raising of indigenous peoples' organizations is a crucial matter. Ultimately, it has to be indigenous peoples who will lead the way towards how the convention may be best implemented. Strengthened organizational capacity of indigenous peoples' institutions and organizations will not only help keep both the ILO and the Government 'on their toes', but also help strengthen alliances and networks with others that are involved in protecting and promoting indigenous peoples' rights. The UNDP-CHTDF project has a component on Capacity Building that human rights activists would do well to monitor and support. Similar activities for the plains Adivasis need to be encouraged.

NETWORKING, ALLIANCES AND SYNERGY-RAISING

All possible actors concerned, indigenous peoples, governments, relevant UN bodies, employers' associations, trades unions, NGOs, human rights organizations and other civic bodies need to coordinate their activities to share responsibilities to promote indigenous peoples' rights. Indigenous peoples have to proactively cultivate effective alliances, including with ILO member states, employers' associations and trades unions, human rights activists, NGOs, and development agencies. Enhanced coordination of the ILO country offices with relevant programmes of other governmental, non-governmental and UN agencies is desirable. Above all, the views of the peoples concerned should be respected, to promote joint efforts, in the spirit of the theme of the *Second International Decade for Indigenous People: Partnership for Action and Dignity*.

COMPLEMENTARY WORK IN HUMAN RIGHTS TREATY BODIES AND OTHER INTERNATIONAL & REGIONAL HUMAN RIGHTS PROCESSES

Human rights advocates and others seeking to implement of the provisions of ILO conventions (such as ILO Conventions No. 107 and 169, or other human rights treaties) should, in addition to working through the ILO supervisory system, explore all possible avenues of redress, particularly the human rights treaty mechanisms offered by the United Nations system (ICCCPR, ICESCR, ICERD, CEDAW, CRC, CAT), the Special Rapporteur mechanisms, the Office of the High Commissioner for Human Rights, the Human Rights Council, the Expert Mechanism on the Rights of Indigenous Peoples, and national and regional human rights commissions and other mechanisms (where they so exist).

RATIFICATION OF ILO CONVENTION NO 169

Ratification of Convention No. 169 by Bangladesh could lead to a marked improvement in the human rights situation of indigenous peoples. This has been recommended in the Poverty Reduction Strategy Paper ("PRSP-II) published in 2008. This was also a recommendation of several member countries of the Human Rights Council during the Universal Periodic Review of members of the Council, including Bangladesh, at the Council's session in February 2009. The recent ratification by Nepal and the supportive role of the ILO's special project on promotion of Convention 169 ("Pro-169") could provide valuable lessons in this regard.

CONSTITUTIONAL REFORM

Constitutional reform for recognition of the indigenous peoples of Bangladesh as indigenous in English and as Adivasi in Bengali, along with constitutional recognition of the special administrative status of the CHT and constitutional entrenchment of the CHT

Accord of 1997 may improve the overall situation. Reforms in this regard would provide a clear direction to policy-makers and government officials alike to eradicate discrimination and provide affirmative action measures, as appropriate. The present Awami League-led Grand Alliance having a two-thirds majority in parliament is an ideal opportunity to lobby for the desired changes.

LEGISLATION

In addition to constitutional reform, legislation is required in Bangladesh to bring several Bangladeshi laws in conformity with the provisions of ILO Convention No. 107 and other human rights treaties ratified by Bangladesh. In this respect, the most important CHT laws requiring reform are the Land Commission Act of 2001 and the CHT Regulation of 1900. The plains require more substantive legislative reforms, the most desirable of which are: (i) the Union Council laws (to reserve seats for Adivasis); (ii) EBSAT Act, 1950 (to include restitution for cases of loss of possession and not just transfer of title and to recognise Adivasi land rights based on customs, usage and prescription); (iii) the Forest Act, 1927 (to recognise the traditional and other rights of inhabitants of reserved, acquired, vested and protected forests and adjacent areas; this applies to an extent to indigenous communities in the CHT as well); (iv) formal recognition of the customary personal laws and juridical systems of the Adivasi peoples.

Bibliography

Adnan, Shapan, *Migration, Land Alienation and Ethnic Conflict: Causes of Poverty in the Chittagong Hill Tracts of Bangladesh*, Research & Advisory Services, Dhaka, 2004.

Barkat Abul, Shafique uz Zaman, Azizur Rahman & Avijit Poddar: *Political Economy of the Vested Property Act in Rural Bangladesh*, Association for Land Reform and Development (ALRD), Dhaka, 1997.

Brauns, Claus-Dieter and Lorenz G. Loffler: *Mru: Hill People on the Border of Bangladesh*, Birkhauser Verlag, Basel, Boston, Berlin, 1990.

Chakraborty, P.: *The Inner-Line Regulation of the NorthEast*, Linkman Publication, India, 1995.

Dansk AV Produktion: *Making a Difference* (Director: Mette Sejsbo, Producer: Anders D. Christensen, camera & Sound: Adam Jeppesen, Music: Troels Oxenvad, Editors: Rikke Malene Nielsen & Bettina Tvede), Copenhagen, 2001.

Hutchinson, R. H. S.: *Chittagong Hill Tracts*, Vivek Publishing House, Delhi, 1978.

Khan, Borhan Uddin: *CHT Governance Study: An Institutional Review*, CARE-Bangladesh, Dhaka, 2004.

Lewin, Capt. T. H.: *The Hill Tracts of Chittagong and the Dwellers Therein; with Comparative vocabularies of the Hill Dialects*, Bengal Printing Company Ltd., Calcutta, 1869.

Mankin, Albert: *The Rights of Adivasis on Land and Forests: Bangladesh Government's Perspectives*, Centre for Indigenous Peoples Research and development (CIPRAD), Dhaka, 2004.

Qureshi, M. S.: *Tribal Cultures of* Bangladesh, Rajshahi University, Rajshahi, 1984.

Roy, Raja Tridiv: *The Departed Melody (Memoirs)*, PPA Publications, Islamabad, 2003.

Schendel, Willem Van : *Francis Buchanan in Southeast Bengal (1798)*, University Press Limited, Dhaka, 1992.

Tomei, Manuela: *Indigenous and Tribal Peoples: An Ethnic Audit of Selected Poverty Reduction Strategy Papers*, International Labour Office, Geneva, 2005.

Index

□□□